Ben Stacy Jerrik (Ed.)

Vipera Ammodytes Gregorwallneri

AF386141

Ben Stacy Jerrik (Ed.)

Vipera Ammodytes Gregorwallneri

List Of Viperine Species And Subspecies, Vipera Ammodytes, Viperinae

Part Press

Imprint

Permission is granted to copy, distribute and/or modify this document under the terms of the GNU Free Documentation License, Version 1.2 or any later version published by the Free Software Foundation; with no Invariant Sections, with the Front-Cover Texts, and with the Back- Cover Texts. A copy of the license is included in the section entitled "GNU Free Documentation License".

All parts of this book are extracted from Wikipedia, the free encyclopedia (www.wikipedia.org).

You can get detailed informations about the authors of this collection of articles at the end of this book. The editors (Ed.) of this book are no authors. They have not modified or extended the original texts.

Pictures published in this book can be under different licences than the GNU Free Documentation License. You can get detailed informations about the authors and licences of pictures at the end of this book.

The content of this book was generated collaboratively by volunteers. Please be advised that nothing found here has necessarily been reviewed by people with the expertise required to provide you with complete, accurate or reliable information. Some information in this book maybe misleading or wrong. The Publisher does not guarantee the validity of the information found here. If you need specific advice (f.e. in fields of medical, legal, financial, or risk management questions) please contact a professional who is licensed or knowledgeable in that area.

Any brand names and product names mentioned in this book are subject to trademark, brand or patent protection and are trademarks or registered trademarks of their respective holders. The use of brand names, product names, common names, trade names, product descriptions etc. even without a particular marking in this works is in no way to be construed to mean that such names may be regarded as unrestricted in respect of trademark and brand protection legislation and could thus be used by anyone.

Cover image: www.ingimage.com
Concerning the licence of the cover image please contact ingimage.

Publisher:
Part Press is a trademark of
International Book Market Service Ltd., 17 Rue Meldrum, Beau Bassin, 1713-01 Mauritius
Email: info@bookmarketservice.com
Website: www.bookmarketservice.com

Published in 2011

Printed in: U.S.A., U.K., Germany. This book was not produced in Mauritius.

ISBN: 978-613-8-96030-0

Contents

Vipera_ammodytes_gregorwallneri

Vipera ammodytes gregorwallneri	
Scientific classification	
Kingdom:	Animalia
Phylum:	Chordata
Subphylum:	Vertebrata
Class:	Reptilia
Order:	Squamata
Suborder:	Serpentes
Family:	Viperidae
Subfamily:	Viperinae
Genus:	*Vipera*
Species:	*V. ammodytes*
Subspecies:	***V. a. gregorwallneri***
Trinomial name	
Vipera ammodytes gregorwallneri Sochurek, 1974	
Synonyms	
• *Vipera ammodytes gregorwallneri* - Sochurek, 1974[1]	

Common names: *(none).*

Vipera ammodytes gregorwallneri is a of venomous viper subspecies [2] found mainly in Austria and the former Yugoslavia.[3]

Geographic range

Found in Austria and the former Yugoslavia.[3]

See also

- List of viperine species and subspecies
- Viperinae by common name
- Viperinae by taxonomic synonyms
- Snakebite

Taxonomy

Many authors, such as Golay *et al.* (1993) do not recognize this taxon and instead relegate it to the synonymy of *V. a. ammodytes*.[4]

References

[1] McDiarmid RW, Campbell JA, Touré T. 1999. Snake Species of the World: A Taxonomic and Geographic Reference, vol. 1. Herpetologists' League. 511 pp. ISBN 1-893777-00-6 (series). ISBN 1-893777-01-4 (volume).

[2] *"Vipera ammodytes gregorwallneri"* (http://www.itis.gov/servlet/SingleRpt/SingleRpt?search_topic=TSN&search_value=635294). Integrated Taxonomic Information System. . Retrieved 10 August 2006.

[3] Strugariu A. 2006. The European Horn-Nosed Viper (http://www.venomousreptiles.org/articles/268). VenomousReptiles.org.

[4] *Vipera ammodytes* (http://reptile-database.reptarium.cz/species.php?genus=Vipera&species=ammodytes) at the Reptarium.cz Reptile Database (http://reptile-database.reptarium.cz/). Accessed 3 August 2007.

List_of_viperine_species_and_subspecies

Viperinae

Asp viper, *Vipera aspis*

Scientific classification	
Kingdom:	Animalia
Phylum:	Chordata
Subphylum:	Vertebrata
Class:	Reptilia
Order:	Squamata
Suborder:	Serpentes
Family:	Viperidae
Subfamily:	**Viperinae** Oppel, 1811

This is a list of all genera, species and subspecies of the subfamily **Viperinae**, otherwise referred to as viperines, true vipers, pitless vipers or Old World vipers. It follows the taxonomy of McDiarmid et al. (1999)[1] and ITIS.[2] For a full list of common names, see Viperinae by common name.

- *Adenorhinos*, Uzungwe mountain bush viper
  - *Adenorhinos barbouri*, Uzungwe mountain bush viper
- *Atheris*, Bush vipers
  - *Atheris anisolepis*
  - *Atheris ceratophora*, Horned bush viper
  - *Atheris chlorechis*, Western bush viper
  - *Atheris desaixi*, Mount Kenya bush viper
  - *Atheris hispida*, Bristly bush viper
  - *Atheris katangensis*, Katanga mountain bush viper
  - *Atheris nitschei*, Great Lakes bush viper
    - *Atheris nitschei nitschei*, Great Lakes bush viper
    - *Atheris nitschei rungweensis*, Rungwe tree viper
  - *Atheris squamigera*, Rough-scaled bush viper
- *Bitis*, Puff adders
  - *Bitis arietans*, Common puff adder
    - *Bitis arietans arietans*, Common puff adder

- *Bitis arietans somalica*, Somali puff adder
- *Bitis atropos*, Mountain adder
- *Bitis caudalis*, Horned adder
- *Bitis cornuta*, Many-horned adder
  - *Bitis cornuta albanica*, Eastern many-horned adder
  - *Bitis cornuta cornuta*, Western many-horned adder
- *Bitis gabonica*, Gaboon viper (3003 Viper)
  - *Bitis gabonica gabonica*, East African gaboon viper
  - *Bitis gabonica rhinoceros*, West African gaboon viper
- *Bitis heraldica*, Angolan adder
- *Bitis inornata*, Plain mountain adder
- *Bitis nasicornis*, Rhinoceros viper
- *Bitis parviocula*, Ethiopian viper
- *Bitis peringueyi*, Peringuey's desert adder
- *Bitis rubida*, Red adder
- *Bitis schneideri*, Namaqua dwarf adder
- *Bitis worthingtoni*, Kenyan horned viper
- *Bitis xeropaga*, Desert mountain adder
- *Cerastes*, Horned vipers
  - *Cerastes Boehmeii* Tunesian horned viper
  - *Cerastes cerastes*, Saharan horned viper
  - *Cerastes gasperettii*, Arabian horned viper
  - *Cerastes vipera*, Sahara sand viper
- *Daboia*, Russell's viper
  - *Daboia russelii*, Russell's viper
    - *Daboia russelii russelii*, Indian Russell's viper
    - *Daboia russelii siamensis*, Eastern Russell's viper
- *Echis*, Saw-scaled vipers
  - *Echis carinatus*, Saw-scaled viper
    - *Echis carinatus astolae*, Astola saw-scaled viper
    - *Echis carinatus carinatus*, South Indian saw-scaled viper
    - *Echis carinatus multisquamatus*, Central Asian saw-scaled viper
    - *Echis carinatus sinhaleyus*, Sri Lankan saw-scaled viper
    - *Echis carinatus sochureki*, Sochurek's saw-scaled viper
  - *Echis coloratus*, Painted saw-scaled viper
  - *Echis hughesi*, Hughes' saw-scaled viper
  - *Echis jogeri*, Joger's saw-scaled viper
  - *Echis leucogaster*, White-bellied saw-scaled viper
  - *Echis megalocephalus*, Cherlin's saw-scaled viper
  - *Echis ocellatus*, West African saw-scaled viper
  - *Echis pyramidum*, Egyptian saw-scaled viper
    - *Echis pyramidum aliaborri*, Red carpet viper
    - *Echis pyramidum leakeyi*, Kenyan carpet viper
    - *Echis pyramidum pyramidum*, Egyptian saw-scaled viper
- *Eristicophis*, McMahon's desert viper

- *Eristicophis macmahonii*, McMahon's desert viper
- *Macrovipera*, Large Palearctic vipers

  - *Macrovipera deserti*, Desert viper
  - *Macrovipera lebetina*, Blunt-nosed viper

    - *Macrovipera lebetina cernovi*
    - *Macrovipera lebetina lebetina*, Blunt-nosed viper
    - *Macrovipera lebetina obtusa*, West-Asian blunt-nosed viper
    - *Macrovipera lebetina transmediterranea*
    - *Macrovipera lebetina turanica*, Turan blunt-nosed viper
  - *Macrovipera mauritanica*, Moorish viper
  - *Macrovipera schweizeri*, Milos viper
  - *Montatheris hindii*, Montane viper
- *Proatheris*, Lowland swamp viper

  - *Proatheris superciliaris*, Lowland swamp viper
- *Pseudocerastes*, Persian horned viper

  - *Pseudocerastes persicus*, Persian horned viper

    - *Pseudocerastes persicus fieldi*, Field's horned viper
- *Vipera*, Palearctic vipers

  - *Montivipera bornmuelleri*, Lebanon viper (Endangered)
  - *Montivipera bulgardaghica*, Mount Bulgar viper
  - *Vipera albicornuta*, Iranian mountain viper
  - *Vipera albizona*, Central Turkish mountain viper
  - *Vipera ammodytes*, Sand viper

    - *Vipera ammodytes ammodytes*, Western sand viper
    - *Vipera ammodytes gregorwallneri*
    - *Vipera ammodytes meridionalis*, Eastern sand viper
    - *Vipera ammodytes montandoni*, Transdanubian sand viper
    - *Vipera ammodytes transcaucasiana*, Transcaucasian sand viper
  - *Vipera anatolica*, Anatolian viper (Critically endangered)
  - *Vipera aspis*, Asp viper

    - *Vipera aspis aspis*, European asp
    - *Vipera aspis atra*, Black asp
    - *Vipera aspis francisciredi*, Central Italian asp
    - *Vipera aspis hugyi*, Southern Italian asp
    - *Vipera aspis zinnikeri*, Gascony asp
  - *Vipera barani*, Baran's adder
  - *Vipera berus*,common viper, Adder

    - *Vipera berus berus*, Common European adder
    - *Vipera berus bosniensis*, Balkan cross adder
    - *Vipera berus sachalinensis*, Sakhalin Island adder
  - *Vipera bornmuelleri*, Bornmuellers viper
  - *Vipera darevskii*, Darevsky's viper
  - *Vipera dinniki*, Dinnik's viper
  - *Vipera ebneri*, Iranian Mountain Steppe viper
  - *Vipera eriwanensis*Armenian Steppe viper

- *Vipera kaznakovi*, Caucasus viper
- *Vipera latastei*, Lataste's viper
  - *Vipera latastei gaditana*
  - *Vipera latastei latastei*, Lataste's viper
- *Vipera latifii*, Latifi's viper
- *Vipera lotievi*, Caucasian meadow viper
- *Vipera magnifica* Magnificent viper
- *Vipera monticola*, Atlas mountain viper
- *Vipera nikolskii*, Nikolski's viper
- *Vipera orlovi* ,Orlov's viper (critically Endangered)
- *Vipera palaestinae*, Palestine viper
- *Vipera pontica*, Pontic adder,Black sea viper
- *Vipera raddei*, Rock viper
  - *Vipera raddei kurdistanica*,Kurdistan viper
- *Vipera renardi* Steppe viper
  - *Vipera renardi renardi*
  - *Vipera renardi parursini*
- *Vipera renardi tienshanica Kasackstan Steppe viper*
- *Vipera seoanei*, Baskian viper
  - *Vipera seoanei cantabrica*
  - *Vipera seoanei seoanei*, Baskian viper
- *Vipera transscaucasicana*, Transcaucasian longnose viper
- *Vipera ursinii*complex, Meadow vipers
  - *Vipera ursinii graeca*, Greek meadow viper
  - *Vipera ursinii macrops*, Albanian meadow viper
  - *Vipera ursinii moldavica*,Romanian meadow viper
  - *Vipera ursinii rakosiensis*,Hungarian meadow viper
  - *Vipera ursinii wettsteini*,Provence meadow viper
  - *Vipera ursinii ursini*, Italian meadow viper
- *Vipera wagneri*, Ocellated mountain viper
- *Vipera xanthina*, Rock viper

See also

- Viperinae by common name
- Viperinae by taxonomic synonyms
- Snakebite

References

[1] McDiarmid RW, Campbell JA, Touré T. 1999. Snake Species of the World: A Taxonomic and Geographic Reference, vol. 1. Herpetologists' League. 511 pp. ISBN 1-893777-00-6 (series). ISBN 1-893777-01-4 (volume).

[2] "Viperinae" (http://www.itis.gov/servlet/SingleRpt/SingleRpt?search_topic=TSN&search_value=563898). Integrated Taxonomic Information System. . Retrieved 23 March 2007.

Vipera ammodytes

Vipera ammodytes	

Scientific classification	
Kingdom:	Animalia
Phylum:	Chordata
Subphylum:	Vertebrata
Class:	Reptilia
Order:	Squamata
Suborder:	Serpentes
Family:	Viperidae
Subfamily:	Viperinae
Genus:	*Vipera*
Species:	***V. ammodytes***
Binomial name	
Vipera ammodytes (Linnaeus, 1758)	
Synonyms	

- *Coluber Ammodytes* - Linnaeus, 1758
- *Vipera Illyrica* - Laurenti, 1768
- *Vipera ammodytes* - Sonnini & Latreille, 1801
- [*Vipera (Echidna)*] *Ammodytes* - Merrem, 1820
- *Cobra ammodytes* - Fitzinger, 1826
- [*Pelias*] *Col*[*uber*]. *ammodytes* - Boie, 1827
- *Vipera (Rhinechis) Ammodytes* - Fitzinger, 1843
- *V*[*ipera*]. (*Vipera*) *ammodytes* - Jan, 1863
- *Vipera ammodytes* - Eber, 1863
- *Vipera ammodytes* - Boulenger, 1896
- [*Vipera ammodytes*] var. *steindachneri* - Werner, 1897
- *Vipera ammodytes* [*ammodytes*] - Boulenger, 1903
- *Vipera ammodytes ammodytes* - Zarevsky, 1915
- *Teleovipera ammodytes* - Reuss, 1927
- *Vipera ammodytes ammodytes* - Mertens & Müller, 1928
- *Rhinaspis illyrica litoralis* - Reuss, 1935
- *Rhinaspis illyrica velebitensis* - Reuss, 1935
- *Rhinaspis illyrica* f[orma]. *melanura* - Reuss, 1937
- *Vipera ammodytes ruffoi* - Bruno, 1968
- *Vipera (Rhinaspis) ammodytes ammodytes* - Obst, 1983

> • *Vipera ammodytes* - Golay *et al.*, 1993[1]

Common names: *horned viper, long-nosed viper, nose-horned viper, sand viper,*[2] *more.*

Vipera ammodytes is a venomous viper species found in southern Europe through to the Balkans and parts of the Middle East. It is reputed to be the most dangerous of the European vipers due to its large size, long fangs (up to 13 mm) and high venom toxicity.[3] The specific name is derived from the Greek words *ammos* and *dutes*, meaning "sand" and "burrower" or "diver"; not a very good name for an animal that actually prefers rocky habitats.[4] Five subspecies are currently recognized, including the nominate subspecies described here.[5]

Description

V. ammodytes

Grows to a maximum length of 95 cm (37.62 in), although individuals usually measure less than 85 cm (33.66 in). Females are somewhat smaller than males. Maximum length also depends on race, with northern forms distinctly larger than southern ones.[3] According to Strugariu (2006), the average length is 50–70 cm (20 to 28 in) with reports of specimens over 1 m (40 in) in length. Females are usually larger and more heavily built, although the largest specimens on record are males.[6]

The head is covered in small, irregular scales that are either smooth or only weakly keeled, except for a pair of large supraocular scales that extend beyond the posterior margin of the eye. 10-13 small scaled border the eye and two rows separate the eye from the supralabials. The nasal scale is large, single (rarely divided) and separated from the rostral by a single nasorostral scale. The rostral scale is wider than it is long.[3]

The most distinctive characteristic is a single "horn" on the snout, just above the rostral scale. It consists of 9-17 scales arranged in 2 (rarely 2 or 4) transverse rows.[3] It grows to a length of about 5 mm and is actually soft and flexible. In southern subspecies, the horn sits vertically upright, while in *V. a. ammodytes* it points diagonally forward.[2]

The body is covered with strongly keeled dorsal scales in 21 or 23 rows (rarely 25) mid-body. The scales bordering the ventrals are smooth or weakly keeled. Males have 133-161 ventral scales and 27-46 paired subcaudals. Females have 135-164 and 24-38 respectively. The anal scale is single.[3]

V. ammodytes

The color pattern is different for males and females. In males, the head has irregular dark brown, dark gray or black markings. A thick, black stripe runs from behind the eye to behind the angle of the jaw. The tongue is usually black and the iris has a golden or coppery color. Males have a characteristic dark blotch or V marking on the back of the head that often connects to the dorsal zigzag pattern. The ground color for males varies and includes many different shades of grey, sometimes yellowish or pinkish grey, or yellowish brown. The dorsal zigzag is dark grey or black, the edge of which is sometimes darker. A row of indistinct, dark (occasionally yellowish) spots runs along each side, sometimes joined in a wavy band.[3]

Females have a similar color pattern, except that it is less distinct and contrasting. They usually lack the dark blotch or V marking on the back of the head that the males have. Ground color is variable and tends more towards browns and bronzes, such as grayish brown, reddish brown, copper, "dirty cream", or brick red. The dorsal zigzag is a shade of brown.[3]

Both sexes have a zigzag dorsal stripe set against a lighter background. This pattern is often fragmented. The belly color varies and can be grayish, yellowish brown, or pinkish, "heavily clouded" with dark spots. Sometimes the ventral color is black or bluish gray with white flecks and inclusions edged in white. The chin is lighter in color than the belly. Underneath, the tip of the tail may be yellow, orange, orange-red, red or green. Melanism does occur, but is rare. Juvenile color patterns are about the same as the adults.[3]

Common names

Horned viper, long-nosed viper, nose-horned viper, sand viper,[2] sand adder, common sand adder, common sand viper,[7] sand natter.[8]

Geographic range

Southern Austria, north-eastern Italy, Slovenia, Croatia, Bosnia-Herzegovina, Serbia, Montenegro, Albania, Republic of Macedonia, Greece (including Greek Macedonia and Cyclades), Romania, Bulgaria, Turkey, Georgia and Syria. The type locality is listed as "Oriente." Schwartz (1936) proposed that the type locality be restricted to "Zara" (Zadar, Croatia).[1]

Conservation status

This species is listed as strictly protected (Appendix II) under the Berne Convention.[9] Middle East

Habitat

The common name *sand viper* is misleading, as this species does not occur in really sandy areas.[10] Mainly, it inhabits dry, rocky hillsides with sparse vegetation. Not usually associated with woodlands, but if so it will be found there around the edges and in clearings. Sometimes found in areas of human habitation, such as railway embankments, farmland, and especially vineyards if rubble piles and stone walls are present. May be found above

2000 m at lower latitudes.[3]

Behaviour

This species has no particular preference for its daily activity period. At higher altitudes, it is more active during the day. At lower altitudes, it may be found at any time of the day, becoming increasingly nocturnal as daytime temperatures rise.[3]

Despite its reputation, this species is generally lethargic, not at all aggressive, and tends not to bite without considerable provocation. If surprised, wild specimens may react in a number of different ways. Some remain motionless and hiss loudly, some hiss and then flee, while still others will attempt to bite immediately.[3]

V. ammodytes hibernates in the winter for a period of 2 to 6 months depending on environmental conditions.[6]

Feeding

Primarily feeds on small mammals and birds. Juveniles apparently prefer lizards. Feeding behavior is influenced by prey size. Larger prey are struck, released, tracked and swallowed, while smaller prey is swallowed without using the venom apparatus. Occasionally, other snakes are eaten.[3] There are also reports of cannibalism.[6]

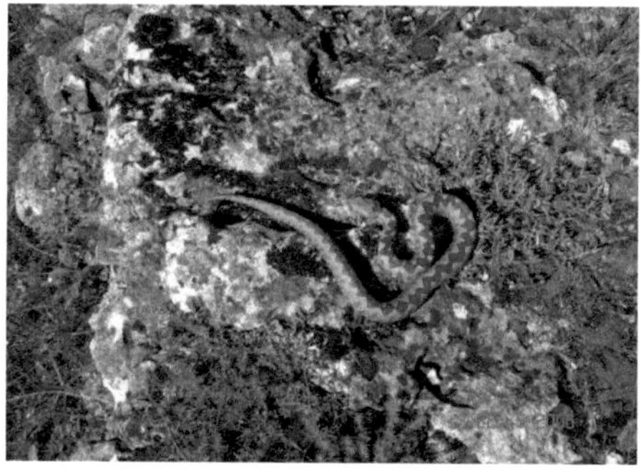

V. ammodytes

Reproduction

Before mating, the males of this species will engage an a combat dance, similar to adders.[3] Mating takes place in the spring (April–May) and between one and twenty live young are born in August–October. At birth, juveniles are 14–24 cm long.[6] This species is ovoviviparous.[11]

Captivity

This species has often been kept in captivity and bred successfully.[3] It tolerates captivity much better than other European vipers, thriving in most surroundings and usually takes food easily from the start.[11] However, as far as handling is concerned, despite its relatively placid reputation, pinning and necking this snake can be risky, as they are relatively strong and can unexpectedly jerk free from a keeper's grasp. For close examinations, it is therefore advisable to use a clear plastic restraining tube instead.[6]

Venom

This is likely the most dangerous snake to be found in Europe. In some areas it is at least a significant medical risk; in the past fatalities were relatively frequent in the Balkans because the peasants there had a habit of walking barefoot.[2]

The venom can be quite toxic [based on tests conducted solely on mice], but varies over time and among different populations.[3] Brown (1973) gives an LD_{50} for mice of 1.2 mg/kg IV, 1.5 mg/kg IP and 2.0 mg/kg SC.[12] Novak *et al.* (1973) give ranges of 0.44-0.82 mg/kg and IV and 0.19-0.64 mg/kg IP. Minton (1974) states 6.6 mg/kg SC.[3]

The venom has both proteolytic and neurotoxic components and contains hemotoxins with blood coagulant properties, similar to and as powerful as in crotalid venom. Other properties include anticoagulant effects, hemoconcentration and hemorrhage. Bites promote symptoms typical of viperid envenomation, such as pain, swelling and discoloration, all of which may be immediate. There are also reports of dizziness and tingling.[3]

Humans respond rapidly to this venom, as do mice and birds. Lizards are less affected, while amphibians may even survive a bite. European snakes, such as *Coronella* and *Natrix*, are possibly immune.[3]

V. ammodytes venom is used in the production of antivenin for the bite of other European vipers and the snake is farmed for this purpose.[7] [11]

Subspecies

Subspecies[5]	Taxon author[5]	Common name	Geographic range
V. a. ammodytes	(Linnaeus, 1758)	Western sand viper[10]	Austria (Styria, Carinthia), north Italy, Slovenia, Croatia, Bosnia-Herzegovina, Serbia, Montenegro, Macedonia, Albania, south-west Romania, north-west Bulgaria[3]
V. a. gregorwallneri	Sochurek, 1974		Austria, the former Yugoslavia[6]
V. a. meridionalis	Boulenger, 1903	Eastern sand viper[10]	Greece (incl. Corfu and other islands), Turkish Thrace[3]
V. a. montandoni	Boulenger, 1904	Transdanubian sand viper[10]	Bulgaria, south Romania[3]
V. a. transcaucasiana	Boulenger, 1913	Transcaucasian sand viper[3]	Georgia, north Turkish Anatolia[3]

Taxonomy

This species was originally described by Linnaeus in *Systema Naturae* in 1758. Subsequently, Boulenger described a number of subspecies in the early 20th century that are still mostly recognized today. However, there are many alternative taxonomies.[3] One additional subspecies that may be encountered in literature is *V. a. ruffoi* (Bruno, 1968),[3] found in the Alpine region of Italy. However, many consider both *ruffoi* and *gregorwalineri* to be synonymous with *V. a. ammodytes*[6] and the taxon *transcaucasiana* to be a separate species.[3] [6]

See also

- List of viperine species and subspecies
- Viperinae by common name
- Viperinae by taxonomic synonyms
- Snakebite

References

[1] McDiarmid RW, Campbell JA, Touré T. 1999. Snake Species of the World: A Taxonomic and Geographic Reference, vol. 1. Herpetologists' League. 511 pp. ISBN 1-893777-00-6 (series). ISBN 1-893777-01-4 (volume).

[2] Street D. 1979. The Reptiles of Northern and Central Europe. London: B.T. Batsford Ltd. 268 pp. ISBN 0-7134-1374-3.

[3] Mallow D, Ludwig D, Nilson G. 2003. True Vipers: Natural History and Toxinology of Old World Vipers. Krieger Publishing Company. 359 pp. ISBN 0-89464-877-2.

[4] Gotch AF. 1986. Reptiles -- Their Latin Names Explained. Poole, UK: Blandford Press. 176 pp. ISBN 0-7137-1704-1.

[5] "Vipera ammodytes" (http://www.itis.gov/servlet/SingleRpt/SingleRpt?search_topic=TSN&search_value=634985). Integrated Taxonomic Information System. . Retrieved 26 July 2006.

[6] Strugariu A. 2006. The European Horn-Nosed Viper (http://www.venomousreptiles.org/articles/268). VenomousReptiles.org.

[7] Mehrtens JM. 1987. Living Snakes of the World in Color. New York: Sterling Publishers. 480 pp. ISBN 0-8069-6460-X.

[8] U.S. Navy. 1991. Poisonous Snakes of the World. US Govt. New York: Dover Publications Inc. 203 pp. ISBN 0-486-26629-X.

[9] Convention on the Conservation of European Wildlife and Natural Habitats, Appendix II (http://conventions.coe.int/treaty/FR/Treaties/Html/104-2.htm) at Council of Europe (http://conventions.coe.int/). Accessed 9 October 2006.

[10] Steward JW. 1971. The Snakes of Europe. Cranbury, New Jersey: Associated University Press (Fairleigh Dickinson University Press). 238 pp. LCCCN 77-163307. ISBN 0-8386-1023-4.

[11] Stidworthy J. 1974. Snakes of the world. Grosset & Dunlap Inc. ISBN 0-448-11856-4.

[12] Brown JH. 1973. Toxicology and Pharmacology of Venoms from Poisonous Snakes. Springfield, Illinois: Charles C. Thomas. 184 pp. LCCCN 73-229. ISBN 0-398-02808-7.

Further reading

- Arnold EN, Burton JA. 1978. A Field Guide to the Reptiles and Amphibians of Britain and Europe. London: Collins. 156 pp. ISBN 3-490-00318-7.

- Biella H-J. 1983. Die Sandotter. Die Neue Brehm-Bücherei. A Ziemsen Verlag. Wittenberg Lutherstadt. 84 pp.

- Bruno S. 1968. Sulla *Vipera ammodytes* in Italia. Memorie del Museo Civico di Storia Naturale, Verona, 15:289-386.

- Copley A, Banerjee S, Devi A. 1973. Studies of snake venom on blood coagulation. Part I: The thromboserpentin (thrombin-like) enzyme in the venoms. Thromb Res 2:487-508.

- Gulden J. 1988. Hibernation and breeding of *V. ammodytes ammodytes*. Litteratura Serpentium 8:168-72.

- Hays WST, Conant Sheila. 2007. Biology and Impacts of Pacific Island Invasive Species. 1. A Worldwide Review of Effects of the Small Indian Mongoose, Herpestes javanicus (Carnivora: Herpestidae) Pacific Science - Volume 61, Number 1, pp. 3–16.

- Nikolsky AM. 1916. Fauna of Russia and adjacent countries. Volume II: Ophidia. Petrograd. Translation from the Israel Program for Scientific Translations, Jerusalem, 1964, 247 pp.

- Mehrtens JM. 1987. Living Snakes of the World in Color. New York: Sterling Publishers. 256 pp. ISBN 0-8069-6460-X.

- Meier J, Stocker KF. 1991. Snake venom protein C activators. In: Tu A, editor. Reptile venoms and Toxins. New York: Marcel Dekker. pp 265–79.

- Mertens R, Wermuth H. 1960. Die Amphibien und Reptilien Europas. Verlag Waldemar Kramer, Frankfurt am Main, 1-264.

- McMahon M. 1990. *Vipera ammodytes meridonalis* envenomation. Journal of the Herpetological Association of Africa, 37:60.

- Petkovic D, Javanovic T, Micevic D, Unkovic-Cvetkovic N, Cvetkovic M. 1979. Action of *Vipera ammodytes* venom and its fractionation on the isolated rat heart. Toxicon, Great Britain, 17:639-44.

- Steward JW. 1971. The Snakes of Europe. London: David & Charles, Newton Abbot. 191 pp. ISBN 0-8386-1023-4.

- U.S. Navy. 1991. Poisonous Snakes of the World. New York: Dover Books. (Reprint of US Govt. Printing Office, Washington D.C.) 133 pp. ISBN 0-486-26629-X.

External links

- *Vipera ammodytes* (http://reptile-database.reptarium.cz/species.php?genus=Vipera&species=ammodytes) at the Reptarium.cz Reptile Database (http://reptile-database.reptarium.cz/). Accessed 21 November 2007.
- *Vipera ammodytes* (http://www.herp.it/indexjs.htm?SpeciesPages/ViperAmmod.htm) at Amphibians and Reptiles of Europe (http://www.herp.it/). Accessed 7 October 2006.
- *Vipera ammodytes* - nose-horned viper (http://bioge.ubbcluj.ro/vivariu/ammodytes_entxt.htm) at Faculty of Biology and Geology (http://bioge.ubbcluj.ro/), "Babes-Bolyai" University, Romania. Accessed 7 October 2006.
- *Vipera ammodytes* (http://www.club100.net/species/V_ammodytes/V_ammodytes.html) at Club100 (http://www.club100.net/). Accessed 7 October 2006.
- *Vipera ammodytes* (http://www.geocities.com/herpetology_bg/Photo_docs/Ph_V_ammodytes.htm) at Amphibians & Reptiles in Bulgaria and Balkan Peninsula (http://web.archive.org/web/20091026235743/http://geocities.com/herpetology_bg/). Accessed 7 October 2006.
- *Vipera ammodytes* mating movie (http://www.vipersgarden.at/movies/mating1/Mating1.html) at VipersGarden (http://www.vipersgarden.at/). Accessed 7 October 2006.
- Video footage of three species, incl. *V. ammodytes*. (https://www.youtube.com/watch?v=2W0hOQk_PPI) on YouTube Accessed 4 November 2006.
- *Vipera ammodytes* giving birth in terrarium. (http://www.mypix.se/user_album.php?us_id=1772&al_id=2953) pictures provided by Terrarium Morbidum (http://www.terrariummorbidum.se/). Accessed 4 June 2009.

Cerastes cerastes

<table>
<tr><td colspan="2" align="center">***Cerastes cerastes***</td></tr>
<tr><td colspan="2" align="center"></td></tr>
<tr><td colspan="2" align="center">**Scientific classification**</td></tr>
<tr><td>Kingdom:</td><td>Animalia</td></tr>
<tr><td>Phylum:</td><td>Chordata</td></tr>
<tr><td>Subphylum:</td><td>Vertebrata</td></tr>
<tr><td>Class:</td><td>Reptilia</td></tr>
<tr><td>Order:</td><td>Squamata</td></tr>
<tr><td>Suborder:</td><td>Serpentes</td></tr>
<tr><td>Family:</td><td>Viperidae</td></tr>
<tr><td>Subfamily:</td><td>Viperinae</td></tr>
<tr><td>Genus:</td><td>*Cerastes*</td></tr>
<tr><td>Species:</td><td>*C. cerastes*</td></tr>
<tr><td colspan="2" align="center">**Binomial name**</td></tr>
<tr><td colspan="2" align="center">***Cerastes cerastes***
(Linnaeus, 1758)</td></tr>
<tr><td colspan="2" align="center">**Synonyms**</td></tr>
<tr><td colspan="2">

- [*Coluber*] *Cerastes* - Linnaeus, 1758
- *Coluber cornutus* - Linnaeus *In* Hasselquist, 1762
- *Cerastes cornutus* - Forskål, 1775
- *Vipera Cerastes* - Sonnini & Latreille, 1801
- *Cerastes Hasselquistii* - Gray, 1842
- *Cerastes Aegyptiacus* - Duméril, Bibron & Duméril, 1854
- *Echidna atricaudata* - Duméril, Bibron & Duméril, 1854
- *Vipera Avicennae* - Jan, 1859
- *V*[*ipera*]. (*Echidna*) *Avicennae* - Jan, 1863
- *V*[*ipera*]. (*Cerastes*) *cerastes* - Jan, 1863
- *Cerastes cornutus* - Boulenger, 1891
- *Cerastes cerastes* - Anderson, 1899
- *Cerastes cornutus* var. *mutila* - Doumergue, 1901
- *Aspis cerastes* - Parker, 1938
- *Cerastes cerastes cerastes* - Leviton & Anderson, 1967
- *Cerastes cerastes* - Werner, Le Verdier, Rosenman & Sivan, 1991
- *Cerastes cerastes karlhartli* - Souchrek, 1974

</td></tr>
</table>

> - *Cerastes cerastes karlhartli* - Tiedemann & Häupl, 1980
> - [*Cerastes cerastes*] *mutila* - Le Berre, 1989
> - *Cerastes cerastes* - Werner, Le Verdier, Rosenman & Sivan, 1991[1]

Common names: *Saharan horned viper,*[2] *horned desert viper,*[3] *more.*

Cerastes cerastes is a venomous viper species native to the deserts of Northern Africa and parts of the Middle East. They often are easily recognized by the presence of a pair of supraocular horns, although hornless individuals do occur.[2] No subspecies are currently recognized.[4]

Description

C. cerastes, with horns.

The average length is 30–60 cm (0.98–2.0 ft), with a maximum of 85 cm (33 in). Females are larger than males.[2]

One of the most distinctive characteristics of this species are the supraorbital horns, one over each eye. However, these may be reduced in size or absent (see *Cerastes*).[2] The eyes are prominent and set on the side of the head.[5] There is significant sexual dimorphism, with males having larger heads and larger eyes than females. Compared to *C. gasperettii*, the relative head size of *C. cerastes* is larger and there is a greater frequency of horned individuals (13% versus 48% respectively).[2] [6]

The color pattern consists of a yellowish, pale gray, pinkish, redish or pale brown ground color that almost always matches the substrate color where the animal is found. Dorsally, a series of dark, semi-rectangular blotches run the length of the body. These may or may not be fused into crossbars. The belly is white and the tail may have a black tip, the tail is usually thin.[2] [5]

Common names

Common names of this species include Saharan horned viper,[2] horned desert viper,[3] Sahara horned viper,[5] desert horned viper, North African horned viper,[7] African desert horned viper, greater cerastes,[8] asp and horned viper.[9] In Egypt it is called *el-ṭorîsha* (حية الطريشة).

Geographic range

Arid north Africa (Morocco, Mauritania and Mali, eastward through Algeria, Tunisia, Niger, Libya and Chad to Egypt, Sudan, Ethiopia and Somalia) through Sinai to the northern Negev of Israel. In the Arabian Peninsula, it occurs in Yemen, extreme southwestern Saudi Arabia and parts of the country in Qatar where it is sympatric with *C. gasperettii*. A report of this species being found in Lebanon is unlikely, according to Joger (1984). Originally, the type locality was listed only as "Oriente." However, Flower (1933) proposed "Egypt" by way of clarification.[1]

Habitat

These snakes favor dry, sandy areas with sparse rock outcroppings, and tend not to prefer coarse sand. Occasionally, they are found around oases, and up to an altitude of 1500 m. Cooler temperatures, with annual averages of 20°C or less, are preferred .[2]

Behavior

They typically move about by sidewinding, during which they press their weight into the sand or soil, leaving whole-body impressions. Often, it is even possible to use these impressions to make ventral scale counts. They have a reasonably placid temperament, but if threatened, they may assume a C-shaped posture and rapidly rub their coils together. Having strongly keeled scales, this produces a rasping noise, similar to *Echis*. They are capable of striking quickly.[2]

Reproduction

C. cerastes, female with eggs.

In captivity, mating was observed in April and always occurred while the animals were buried in the sand.[2] This species is oviparous, laying 8-23 eggs that hatch after 50 to 80 days of incubation. The eggs are laid under rocks and in abandoned rodent burrows. The hatchlings measure 12–15 cm in length.[5]

Venom

C. cerastes venom is not very toxic, although it is reported to be similar in action to *Echis* venom.[2] Envenomation usually causes swelling, hemorrhage, necrosis, nausea, vomiting and hematuria. A high phospholipase A2 content may cause cardiotoxicity and myotoxicity.[5] Studies of venom from both *C. cerastes* and *C. vipera* list a total of eight venom fractions, the most powerful of which has hemorrhagic activity. Venom yields vary, with ranges of 19–27 mg to 100 mg of dried venom being reported.[2] For venom toxicity, Brown (1973) gives LD_{50} values of 0.4 mg/kg IV and 3.0 mg/kg SC.[7] An estimated lethal dose for humans is 40–50 mg.[5]

Taxonomy

A number of subspecies may be encountered in literature:[2]

* *C. c. hoofieni* - Werner & Sivan, 1999 - Saudi Arabia.
* *C. c. karlhartli* - Sochurek, 1974 - Egyptian horned viper - southeast Egypt and Sinai Peninsula.
* *C. c. mutila* - Domergue, 1901 - Algerian horned viper - southwest Algeria, Morocco.

Previously, *C. gasperettii* was also regarded as a subspecies of *C. cerastes*.[2]

See also

- List of viperine species and subspecies
- Viperinae by common name
- Viperinae by taxonomic synonyms
- Snakebite
- Viper (hieroglyph)

References

[1] McDiarmid RW, Campbell JA, Touré T. 1999. Snake Species of the World: A Taxonomic and Geographic Reference, vol. 1. Herpetologists' League. 511 pp. ISBN 1-893777-00-6 (series). ISBN 1-893777-01-4 (volume).

[2] Mallow D, Ludwig D, Nilson G. 2003. True Vipers: Natural History and Toxinology of Old World Vipers. Krieger Publishing Company. 359 pp. ISBN 0-89464-877-2.

[3] Mehrtens JM. 1987. Living Snakes of the World in Color. New York: Sterling Publishers. 480 pp. ISBN 0-8069-6460-X.

[4] *"Cerastes cerastes"* (http://www.itis.gov/servlet/SingleRpt/SingleRpt?search_topic=TSN&search_value=634963). Integrated Taxonomic Information System. . Retrieved 30 July 2006.

[5] Spawls S, Branch B. 1995. The Dangerous Snakes of Africa. Ralph Curtis Books. Dubai: Oriental Press. 192 pp. ISBN 0-88359-029-8.

[6] Werner YL, Verdier A, Rosenman D, Sivan N. 1991. Systematics and zoogeography of Cerastes (Ophidia: Viperidae) in the Levant: 1, Distinguishing Arabian from African "Cerastes cerastes." The Snake 23:90-100.

[7] Brown JH. 1973. Toxicology and Pharmacology of Venoms from Poisonous Snakes. Springfield, Illinois: Charles C. Thomas. 184 pp. LCCCN 73-229. ISBN 0-398-02808-7.

[8] U.S. Navy. 1991. Poisonous Snakes of the World. US Govt. New York: Dover Publications Inc. 203 pp. ISBN 0-486-26629-X.

[9] Ditmars RL. 1933. Reptiles of the World. Revised Edition. The MacMillan Company. 329 pp. 89 plates.

Further reading

- Calmette A. 1907. Les venins, les animaux venimeux et la serotherapie antivenimeuse. In: Bucherl W. editor. 1967. *Venomous Animals and Their Venoms*. Vol. I. Paris: Masson. pp 233.
- Mohamed AH, Kamel A, Ayobe MH. 1969. "Studies of phospholipase A and B activities of Egyptian snake venoms and a scorpion venom". *Toxicon* 6:293-8.
- Joger U. 1984. *The Venomous Snakes of the Near and Middle East*. Wiesbaden: Dr. Ludwig Reichert Verlag. 175 pp.
- Labib RS, Malim HY, Farag NW. 1979. "Fractionation of *Cerastes cerastes* and *Cerastes vipera* snake venoms by gel filtration and identification of some enzymatic and biological activities". *Toxicon* 17:337-45.
- Labib RS, Azab MH, Farag NW. 1981. "Effects of *Cerastes cerastes* (Egyptian sand viper) snake venoms on blood coagulation: separation of coagulant and anticoagulant factors and their correlation with arginineesterase protease activities". *Toxicon* 19:85-94.
- Labib RS, Azab ER, Farag NW. 1981. "Proteases of *Cerastes cerastes* and *Cerastes vipera* snake venoms". *Toxicon* 19:73-83.
- Schneemann M, Cathomas R, Laidlaw ST, El Nahas AM, Theakston RDG, Warrell DA. 2004. "Life-threatening envenoming by the Saharan horned viper (*Cerastes cerastes*) causing micro-angiopathic haemolysis, coagulopathy and acute renal failure: clinical cases and review". *Association of Physicians. QJM* vol. 97 no. 11. Full text (http://qjmed.oxfordjournals.org/cgi/content/full/97/11/717) at Oxford Journals (http://qjmed.oxfordjournals.org/). Accessed 9 March 2007.
- Schnurrenburger H. 1959. "Observations on behavior in two Libyan species of viperine snake". *Herpetologica* 15:70-2.
- U.S. Navy. 1991. *Poisonous Snakes of the World*. New York: Dover Books. (Reprint of US Govt. Printing Office, Washington D.C.) 133 pp. ISBN 0-486-26629-X.

External links

- *Cerastes cerastes* (http://reptile-database.reptarium.cz/species.php?genus=Cerastes&species=cerastes) at the Reptarium.cz Reptile Database (http://reptile-database.reptarium.cz/). Accessed 2 August 2007.
- Sand viper page (http://www.plumed-serpent.com/dscour.html) at Plumed-serpent.com (http://www.plumed-serpent.com/). Accessed 30 July 2006.
- Video of *Cerastes cerastes* (https://www.youtube.com/watch?v=MjOMbVIUJSQ) on YouTube. Accessed 31 May 2007.

Vipera berus

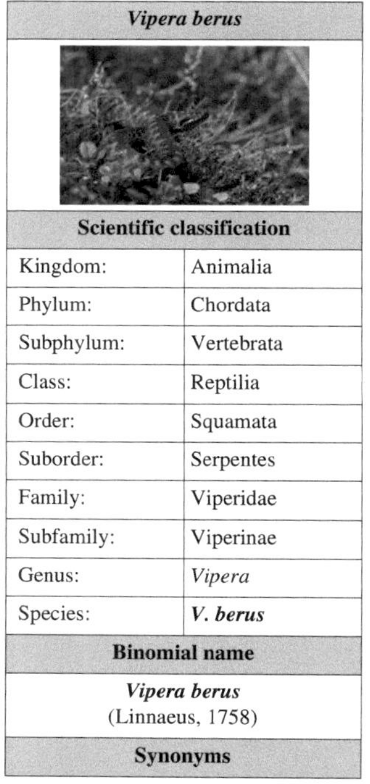

Vipera berus	
Scientific classification	
Kingdom:	Animalia
Phylum:	Chordata
Subphylum:	Vertebrata
Class:	Reptilia
Order:	Squamata
Suborder:	Serpentes
Family:	Viperidae
Subfamily:	Viperinae
Genus:	*Vipera*
Species:	*V. berus*
Binomial name	
Vipera berus (Linnaeus, 1758)	
Synonyms	

Vipera berus, the **common European adder**[2] or **common European viper**,[3] is a venomous viper species that is extremely widespread and can be found throughout most of Western Europe and all the way to Far East Asia.[1] Known by a host of common names including *Common adder* and *Common viper*, adders have been the subject of much folklore in Britain and other European countries.[4] They are not regarded as highly dangerous;[2] the snake is not aggressive and usually only bites when alarmed or disturbed. Bites can be very painful, but are seldom fatal.[5] The specific name, *berus*, is New Latin and was at one time used to refer to a snake, possibly the grass snake, *Natrix natrix*.[6]

The common adder is found in different terrains, habitat complexity being essential for different aspects of its behaviour. It feeds on small mammals, birds, lizards, amphibians and in some cases on spiders, worms and insects. Females breed once every two or three years with litters usually born in late summer to early autumn in the Northern hemisphere. The common adder, like most other vipers, is ovoviviparous; litters range in size from 3 to 20 with young staying with their mothers for a few days. Adults grow to a length of 60 to 90 centimetres (24 to 35 in) and a mass of 50 grams (1.8 oz) to about 180 grams (6.3 oz). Three subspecies are recognized, including the nominate subspecies described here.[7] The snake is not considered to be threatened though it is protected in some countries.

Description

Relatively thick-bodied, Maximum size varies per region. The largest, at over 90 centimetres (35 in), are found in Scandinavia; specimens of 104 centimetres (41 in) have been observed there on two occasions. In France and Great Britain, the maximum size is 80–87 centimetres (31–34 in).[2] Mass ranges from 50 grams (1.8 oz) to about 180 grams (6.3 oz)[8] [9]

V. berus: normal and melanistic colour patterns.

The head is fairly large and distinct, the sides of which are almost flat and vertical. The edge of the snout is usually raised into a low ridge. Seen from above, the rostral scale is not visible, or only just. Immediately behind the rostral, there are 2 (rarely 1) small scales. Dorsally, there are usually 5 large plates: a squarish frontal (longer than wide, sometimes rectangular), 2 parietals (sometimes with a tiny scale between the frontal and the parietals), and 2 long and narrow supraoculars. The latter are large and distinct, each separated from the frontal by 1-4 small scales. The nostril is situated in a shallow depression within a large nasal scale. The eye is relatively large—equal in size or slightly larger than the nasal scale—but often smaller in females. Below the supraoculars there are 6-13 (usually 8-10) small circumorbital scales. The temporal scales are smooth (rarely weakly keeled). There are 10-12 sublabials and 6-10 (usually 8-9) supralabials. Of the latter, the numbers 3 and 4 are the largest, while 4 and 5 (rarely 3 and 4) are separated from the eye by a single row of small scales (sometimes two rows in alpine specimens).[2]

Midbody there are 21 dorsal scales rows (rarely 19, 20, 22, or 23). These are strongly keeled scales, except for those bordering the ventral scales. These scales seem loosely attached to the skin and lower rows become increasingly wide; those closest to the ventral scales are twice as wide as the ones along the midline. The ventral scales number 132-150 in males and 132-158 in females. The anal plate is single. The subcaudals are paired, numbering 32-46 in males and 23-38 in females.[2]

The color pattern varies, ranging from very light-colored specimens with small incomplete dark dorsal crossbars to entirely brown ones with faint or clear darker brown markings, and on to melanistic individuals that are entirely dark and lack any apparent dorsal pattern. However, most have some kind of zigzag dorsal pattern down the entire length of the body and tail. The head usually has a distinctive dark V or X on the back. A dark streak runs from the eye to the neck and continues as a longitudinal series of spots along the flanks.[2] Unusual for snakes, the sexes are possible to tell apart by the colour. Females are usually brownish in hue with dark-brown markings, the males are pure grey with black markings. The basal colour of males will often be a tad lighter than that of the females, making the black zigzag pattern stand out. The melanistic individuals are often females.

Common names

In keeping with its wide distribution and familiarity through the ages, *Vipera berus* has a large number of common names, which include:

Common European adder,[2] *common European viper,*[3] *European viper,*[10] *northern viper,*[11] *adder, common adder, crossed viper, European adder,*[12] *common viper, European common viper, cross adder,*[13] or *common cross adder.*[14]

In Sweden, Norway, and Denmark, it is known as "huggorm" or "hoggorm". In Finland, it is known as "kyykäärme" or simply "kyy." The word "adder" was *nædre* in Old English; in the 14th century *a nadder* was, like *a napron*, rebracketed as *an adder*. It appears with the generic meaning of serpent in the older forms of many Germanic languages, and is thus used in the Old English version of the Christian Scriptures for the devil, the *serpent* of the Book of Genesis.[4] [15]

Geographic range

Vipera berus has a wide range. It can be found across the Eurasian land-mass; from northwestern Europe (Great Britain, Scandinavia, Germany, France) across southern Europe (Italy, Albania, Croatia, Republic of Macedonia, Bulgaria, and northern Greece) and eastern Europe to north of the Arctic Circle, and Russia to the Pacific Ocean, Sakhalin Island, North Korea, northern Mongolia and northern China. The type locality was originally listed as "Europa". Mertens and Müller (1940) proposed restricting the type locality to "Upsala, Schweden" (Uppsala, Sweden)[1] and it was eventually restricted to Berthåga, Uppsala by designation of a neotype by Krecsák & Wahlgren (2008).[16]

In several European countries it is notable as being the only native venomous snake.

Conservation status

In the United Kingdom, it is illegal to kill, injure, harm, or sell adders under the 1981 Wildlife and Countryside Act.[17] The common viper is categorised as "endangered" in Switzerland,[18] and is also protected in some other countries in its range. It is also found in many protected areas.[19] This species is listed as protected (Appendix III) under the Berne Convention.[20]

The International Union for the Conservation of Nature Red List of Threatened Species describes the conservation status as of "least concern" in view of its wide distribution, presumed large population, broad range of habitats, and likely slow rate of decline though it acknowledges the population to be decreasing.[21] Reduction in habitat for a variety of reasons, fragmentation of populations in Europe due to intense agriculture practices, and collection for the pet trade or for venom extraction have been recorded as major contributing factors for its decline.[19]

Habitat

V. berus

Sufficient habitat complexity is a crucial requirement for the presence of this species, in order to support their various behaviors—basking, foraging, and hibernation—as well as to offer some protection from predators and human harassment.[2] It is found in variety of habitats, including: chalky downs, rocky hillsides, moors, sandy heaths, meadows, rough commons, edges of woods, sunny glades and clearings, bushy slopes and hedgerows, dumps, coastal dunes, and stone quarries. They will venture into wetlands if dry ground is available nearby. Therefore, they may be found on the banks of streams, lakes, and ponds.[22]

In much of southern Europe, such as southern France and northern Italy, it is found in either low lying wetlands or at high altitudes. In the Swiss Alps, it may ascend to about 3,000 m (9,842 ft). In Hungary and Russia, it avoids open steppeland; a habitat in which *V. ursinii* is more likely to occur. In Russia, however, it does occur in the forest steppe zone.[22]

Behaviour

This species is mainly diurnal, especially in the north of its range. Further south it is said[23] to be active in the evening, and it may even be active at night during the summer months. It is predominantly a terrestrial

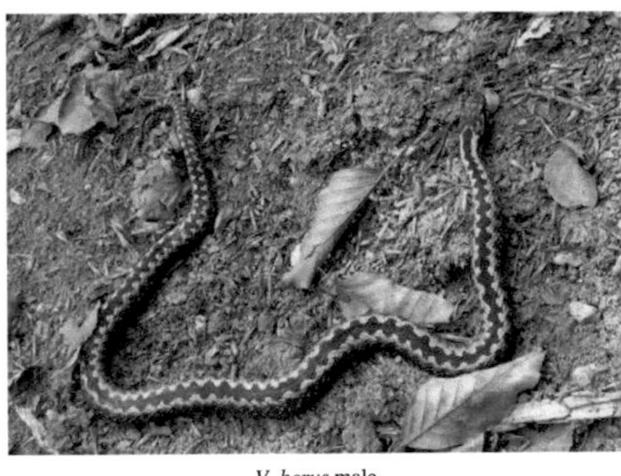

V. berus

species, although it has been known to climb up banks and into low bushes in order to bask or search for prey.[22]

Adders are not usually aggressive, tending to be rather timid and biting only when cornered or alarmed. People are generally only bitten after stepping on them or attempting to pick them up. They will usually disappear into the undergrowth at a hint of any danger, but will return once all is quiet, often to the same spot. Occasionally, individual snakes will reveal their presence with a loud and sustained hissing, hoping to warn off potential aggressors. Often, these turn out to be pregnant females. When threatened, the front part of the body is drawn into an S-shape to prepare for a strike.[22]

V. berus male

The species is cold-adapted and hibernates in the winter. In Great Britain, males and females hibernate for about 150 and 180 days respectively. In northern Sweden hibernation lasts 8–9 months. On mild winter days, they may emerge to bask where the snow has melted and will often travel across snow. About 15% of adults and 30-40% of juveniles die during hibernation.[2]

Feeding

Diet consists mainly of small mammals, such as mice, voles, and shrews, as well as lizards. Sometimes, slow worms are taken, and even weasels and moles. They feed on amphibians, such as frogs, newts, and salamanders. Birds are also reported[24] to be on the menu, especially nestlings and even eggs, for which they will climb into shrubbery and bushes. Generally, diet varies depending on locality.[22] Juveniles will eat nestling mammals, small lizards and frogs as well as worms and spiders. Once they reach about 30 cm (1 ft) in length, their diet begins to resemble that of the adults.[2]

Reproduction

In Hungary, mating takes place in the last week of April, while in the north it happens later in the second week of May. Matings have also been observed in June and even early October, but it is not known if the autumn matings result in any young.[2] Females often breed once every two years,[22] or even once every three years if the seasons are short and the climate is severe.[2]

V. berus - showing strongly keeled scales on dorsal area.

Males find females by following their scent trails, sometimes tracking them for hundreds of meters a day. If a female is found and flees, the male follows. Courtship involves side-by-side parallel "flowing" behavior, tongue flicking along the back and excited lashing of the tail. Pairs stay together for one or two days after mating. Males chase away their rivals and engage in combat. Often, this also starts with the aforementioned flowing behavior before culminating in the dramatic "adder dance."[2] In this act, the males confront each other, raise up the front part of the body vertically, make swaying movements and attempt to push each other to the ground. This is repeated until one of the two becomes exhausted and crawls off to find another mate. Interestingly, Appleby (1971) notes that he has never seen an intruder win one of these contests, as if the frustrated defender is so aroused by courtship that he refuses to lose his chance to mate.[25] There are no records of any biting taking place during these bouts.[22]

Females usually give birth in August–September, but sometimes as early as July, or as late as early October. Litters range in size from 3 to 20. The young are usually born encased in a transparent sac from which they must free themselves. Sometimes, they succeed in freeing themselves from this membrane while still inside the female. The neonates, measuring 14 to 23 cm (average of 17 cm; 7 in), are born with a fully functional venom apparatus and a reserve supply of yolk within their bodies. They shed their skins for the first time within a day or two. Females do not appear to take much interest in their offspring, but the young have been observed to remain near their mothers for several days after birth.[22]

Venom

Because of the rapid rate of human expansion throughout the range of this species, bites are relatively common. Domestic animals and livestock are frequent victims. In Great Britain, most instances occur in March–October. In Sweden, there are about 1,300 bites a year, with an estimated 12% that require hospitalisation.[2] At least eight different antivenoms are available against bites from this species.[26]

Mallow et al. (2003) describe the venom toxicity as being relatively low compared to other viper species. They cite Minton (1974) who reported the LD_{50} values for mice to be 0.55 mg/kg IV, 0.80 mg/kg IP and 6.45 mg/kg SC. As a comparison, in one test the minimum lethal dose of for a guinea pig was 40–67 mg, but only 1.7 mg was necessary when *Daboia russelii* venom was used.[2] Brown (1973) gives a higher subcutaneous LD_{50} range of 1.0-4.0 mg/kg.[14] All agree that the venom yield is low: Minton (1974) mentions 10–18 mg for specimens 48–62 cm (19–24 in) in length,[2] while Brown (1973) lists only 6 mg.[14]

Relatively speaking, bites from this species are not highly dangerous.[2] In Britain there have been only 14 known fatalities since 1876; the last a 5-year-old child in 1975.[5] and one near fatal bite of a 39 year old woman in Essex in 1998[5] . An 82-year-old woman died following a bite in Germany in 2004, although it is not clear whether her death was due to the effect of the venom.[27] Even so, professional medical help should always be sought as soon as possible after any bite. Very occasionally bites can be life threatening, particularly in small children, while adults may experience discomfort and disability long after the bite.[5] The length of recovery varies, but may take up to a year.[2]

Vipera berus; the one erect fang has left a small venom stain on the glove.

Local symptoms include immediate and intense pain, followed after a few minutes (but perhaps by as much as 30 minutes) by swelling and a tingling sensation. Blisters containing blood are not common. The pain may spread within a few hours, along with tenderness and inflammation. Reddish lymphangitic lines and bruising may appear, and the whole limb can become swollen and bruised within 24 hours. Swelling may also spread to the trunk, and with children, throughout the entire body. Necrosis and intracompartmental syndromes are very rare.[5]

Systemic symptoms resulting from anaphylaxis can be dramatic. These may appear within 5 minutes post bite, or can be delayed for many hours. Such symptoms include nausea, retching and vomiting, abdominal colic and diarrhoea, incontinence of urine and faeces, sweating, fever, vasoconstriction, tachycardia, lightheadedness, loss of consciousness, blindness,[28] shock, angioedema of the face, lips, gums, tongue, throat and epiglotis, urticaria and bronchospam. If left untreated, these symptoms may persist or fluctuate for up to 48 hours.[5] In severe cases, cardiovascular failure may occur.[2]

Taxonomy

The species has three recognised subspecies :

Subspecies[7]	Taxon author[7]	Common name	Geographic range
V. b. berus	(Linnaeus, 1758)	Common European adder[2]	Norway, Sweden, Finland, Latvia, Estonia, Lithuania, France, Denmark, Germany, Austria, Switzerland, Northern Italy, Belgium, Netherlands, Great Britain, Poland, Czech Republic, Slovakia, Hungary, Romania, Russia, Mongolia, Northwest China (north Xinjiang)
V. b. bosniensis	Boettger, 1889	Balkan cross adder[13]	Balkan Peninsula
V. b. sachalinensis	Zarevskij, 1917	Sakhalin Island adder[12]	Russian Far East (Amur Oblast, Primorskye Kray, Khabarovsk Kray, Sakhalin Island), North Korea, north-east China (Jilin)

The subspecies *V. b. bosniensis* and *V. b. sachalinensis* have been regarded as full species in some recent publications.[2]

See also

- List of viperine species and subspecies
- Viperinae by common name
- Viperinae by taxonomic synonyms
- List of reptiles of Italy

References

[1] McDiarmid RW, Campbell JA, Touré T. 1999. Snake Species of the World: A Taxonomic and Geographic Reference, vol. 1. Herpetologists' League. 511 pp. ISBN 1-893777-00-6 (series). ISBN 1-893777-01-4 (volume).

[2] Mallow D, Ludwig D, Nilson G. 2003. True Vipers: Natural History and Toxinology of Old World Vipers. Krieger Publishing Company, Malabar, Florida. 359 pp. ISBN 0-89464-877-2.

[3] Stidworthy J. 1974. Snakes of the World. Grosset & Dunlap Inc. 160 pp. ISBN 0-448-11856-4.

[4] ""Everyday adders" - the Adder in Folklore" (http://www.crislis.co.uk/adder/folklore.htm). The Herpetological Conservation Trust. . Retrieved 7 February 2010.

[5] Warrell DA. 2005. Treatment of bites by adders and exotic venomous snakes. British Medical Journal 331:1244-7. PDF (http://www.bmj. com/cgi/reprint/331/7527/1244.pdf) at bmj.com (http://www.bmj.com/). Accessed 15 September 2006.

[6] Gotch AF. 1986. Reptiles -- Their Latin Names Explained. Poole, UK: Blandford Press. 176 pp. ISBN 0-7137-1704-1.

[7] "Vipera berus" (http://www.itis.gov/servlet/SingleRpt/SingleRpt?search_topic=TSN&search_value=634988). Integrated Taxonomic Information System. . Retrieved 15 August 2006.

[8] Mats Olsson, Thomas Madsen and Richard Shine, "Is sperm really so cheap? Costs of reproduction in male adders,*Vipera berus*", *Proceedings of the Royal Society* 1997 264, p 456 (http://rspb.royalsocietypublishing.org/content/264/1380/455.full.pdf) (includes chart showing range of male mass in one population)

[9] Alexandru STRUGARIU, Ştefan R. ZAMFIRESCU and Iulian GHERGHEL "First record of the adder (Vipera berus berus) in Argeş County (Southern Romania)", *Biharean Biologist* (2009), 3, 2, p 164 [(http)://biologie-oradea.xhost.ro/BihBiol/cont/v3n2/bb.031206.Strugariu.pdf](gives example masses of females)

[10] U.S. Navy. 1991. Poisonous Snakes of the World. US Govt. New York: Dover Publications Inc. 203 pp. ISBN 0-486-26629-X.

[11] *Vipera berus* (http://reptile-database.reptarium.cz/species.php?genus=Vipera&species=berus) at the Reptarium.cz Reptile Database (http://reptile-database.reptarium.cz/). Accessed 21 November 2007.

[12] Mehrtens JM. 1987. Living Snakes of the World in Color. New York: Sterling Publishers. 480 pp. ISBN 0-8069-6460-X.

[13] Steward JW. 1971. The Snakes of Europe. Cranbury, New Jersey: Associated University Press (Fairleigh Dickinson University Press). 238 pp. LCCCN 77-163307. ISBN 0-8386-1023-4.

[14] Brown JH. 1973. Toxicology and Pharmacology of Venoms from Poisonous Snakes. Springfield, Illinois: Charles C. Thomas. 184 pp. LCCCN 73-229. ISBN 0-398-02808-7.

[15] Dictionary.com. "adder" (http://dictionary.reference.com/browse/adder). *Dictionary.com Unabridged*. Random House,Inc.. . Retrieved 7 February 2010.

[16] Krecsák L. & Wahlgren R. (2008): A survey of the Linnaean type material of *Coluber berus, Coluber chersea* and *Coluber prester* (Serpentes, Viperidae). *The Journal of Natural History* 42(35–36): 2343–2377. doi:10.1080/00222930802126888

[17] "Adder (Vipera berus)" (http://www.arkive.org/adder/vipera-berus/facts-and-status.html). *Arkive (Images of life on Earth)*. www.wildscreen.org.uk. . Retrieved 7 February 2010.. This ref cites Beebee, T. & Griffiths, R. (2000) *Amphibians and reptiles: a natural*

history of the British herpetofauna. (http://books.google.co.in/books?id=JMF1QgAACAAJ&dq=Amphibians+and+reptiles:+a+
natural+history+of+the+British+herpetofauna&cd=1) Harper Collins Publishers Ltd, London. as the source.

[18] Monney JC, Meyer A. 2005. *Rote Liste der gefährdeten Reptilien der Schweiz.* Hrsg. Bundesamt für Umwelt, Wald und Landschaft
      BUWAL, Bern und Koordinationsstelle für Amphibien- und Reptilienschutz der Schweiz, Bern. BUWAL-Reihe. 50 pp.

[19] "Vipera berus" (http://www.iucnredlist.org/apps/redlist/details/157248/0). *2008-09 IUCN Red List of Threatened Species Version
      2009.2.* www.iucnredlist.org. . Retrieved 7 February 2010.

[20] Convention on the Conservation of European Wildlife and Natural Habitats, Appendix III (http://74.125.153.132/
      search?q=cache:B7MTMDETJ20J:conventions.coe.int/treaty/en/reports/html/185.htm+http://conventions.coe.int/Treaty/FR/
      Treaties/Html/104.htm+cached&cd=1&hl=en&ct=clnk&gl=in&client=firefox-a) at Council of Europe (http://74.125.153.132/
      search?q=cache:Bxx83m59bgwJ:conventions.coe.int/+http://conventions.coe.int/&cd=1&hl=en&ct=clnk&gl=in&client=firefox-a).
      Accessed 7 February 2010.

[21] "IV: The Categories" (http://www.iucnredlist.org/technical-documents/categories-and-criteria/2001-categories-criteria). *2001 IUCN Red
      List Categories and Criteria version 3.1.* www.iucnredlist.org. . Retrieved 14 February 2010.

[22] Street D. 1979. The Reptiles of Northern and Central Europe. London: B.T. Batsford Ltd. ISBN 0-7134-1374-3.

[23] Boulenger, G.A.. (1913).*Snakes of Europe.* Methuen & Co, London (http://www.archive.org/details/snakesofeurope00boul)

[24] Leighton, Gerald R. (1901). *The Life-History of British Serpents and Their Local Distribution in the British Isles* (http://books.google.
      com/?id=7i_8ZmymfMoC). Edinburgh & London: Blackwood & sons. p. 84. ISBN 1444630911. . Retrieved 8 February 2010.

[25] Appleby, L. G. (1971). *British Snakes.* London (Baker), 150 pages. ISBN 0212983938.

[26] *Vipera berus* antivenoms (http://www.toxinfo.org/antivenoms/indication/VIPERA_BERUS.html) at Munich Antivenom Index (http://
      www.toxinfo.org/antivenoms/). Accessed 15 September 2006.

[27] Tod durch Kreuzotterbiss? (http://www.ggiz-erfurt.de/aktuelles/akt_press_04_juli_kreuzotter_ostsee.htm) at Gemeinsames
      Giftinformationszentrum (http://www.ggiz-erfurt.de/). Accessed 25 May 2007.

[28] The Daily Mail (http://www.dailymail.co.uk/news/article-1296520/
      Adder-bite-leaves-father-blinded-choking-just-minutes-death-walk-family.html), *"Adder bite leaves father blinded, choking and just minutes
      from death on walk with family,"* (21 July 2010 - retrieved on 21 July 2010).

Further reading

- Ananjeva NB, Borkin LJ, Darevsky IS, Orlov NL. 1998. Amphibians and Reptiles. Encyclopedia of Nature of
  Russia. ABF Moscow (in Russian). 574 pp.
- Appleby L.G. 1971. British Snakes. London: J. Baker. 201 pp. ISBN 0-212-98393-8.
- Joger U, Lenk P, Baran I, Böme W, Ziegler T, Heidrich P, Wink M. 1997. The phylogenetic position of *Vipera
  barani* and of *Vipera nikolskii* within the *Vipera berus* complex.
- Minton S.A. Jr. 1974. Venom Diseases. Springfield (IL): CC Thomas Publ. 386 pp.
- Wüster W, Allum CSE, Bjargardottir IB, Bailey KL, Dawson KJ, Guenioui J, Lewis J, McGurk J, Moore AG,
  Niskanen M, Pollard CP. 2004. Do aposematism and Batesian mimicry require bright colours? A test, using
  European viper markings. Proceedings of the Royal Society of London. B 271 pp 2495–2499. PDF (http://www.
  bangor.ac.uk/~bss166/Publications/2004_Viper_Aposematism_online.pdf) at Wolfgang Wüster, School of
  Biological Sciences, University of Wales, Bangor (http://www.bangor.ac.uk/~bss166/). Accessed on 15
  August 2006.

External links

- *Vipera berus* (http://www.surrey-arg.org.uk/cgi-bin/SARG2ReptileSpeciesData.asp?Species=Adder) at
  Surrey Amphibian and Reptile Group (SARG) (http://www.surrey-arg.org.uk)
- *Vipera berus* European Field Herping Community (http://www.euroherp.com/species.php?sp=181)
- *Vipera berus* (http://www.herp.it/SpeciesPages/ViperBerus.htm) at Amphibians and Reptiles of Europe
  (http://www.herp.it/). Accessed on 16 August 2006.
- *Vipera berus* (http://www.arkive.org/adder/vipera-berus/) media at ARKive
- Adder or Viper - *Vipera berus* (http://www.herpetofauna.co.uk/adder.htm) at Reptiles and Amphibians of the
  UK (http://www.herpetofauna.co.uk/). Accessed 9 October 2006.
- *Vipera berus* (http://www.club100.net/species/V_berus/V_berus.html) at Club100 (http://www.club100.
  net/). Accessed 9 October 2006.

- *Viper berus* - Adder (http://www.first-nature.com/reptiles/vipera_berus.htm) at First Nature (http://www.first-nature.com/). Accessed 9 October 2006.
- Adder (*Vipera berus*) (http://www.wartsoc.co.uk/gallery/warksR.php) at Warwickshire Amphibian and Reptile Team (http://www.wartsoc.co.uk/herpetofauna.htm). Accessed 11 February 2010 .
- Adder, *Vipera berus* (http://www.herpfrance.com/reptile/adder_vipera_berus.php) at Reptiles & Amphibians of France (http://www.herpfrance.com/). Accessed 6 June 2008.
- *Vipera berus* images (http://en.hribi.net/zivali.asp?id=4) at Hribi.net (http://en.hribi.net/). Accessed 7 February 2010.
- Snakes (http://web.archive.org/web/20080413000721/http://www.froglife.org/speciesIDsReptiles.htm) at Froglife, UK (now part of Amphibian & reptile conservation Trust) (http://www.arc-trust.org/). Dead site archived by www.archive.org. Accessed 11 February 2010.
- Add an Adder (http://www.adder.org.uk/) (UK Herpetological Conservation Trust). Accessed 31 December 2007.
- James Stroud Research and Contact Details (http://www2.hull.ac.uk/scarborough/campus-departments/environmental-and-marine-scien/staff/cems-postgraduate/james-stroud.aspx) (University of Hull, UK)

Viperinae

<table>
<tr><td colspan="2" align="center">True vipers</td></tr>
<tr><td colspan="2" align="center">Asp viper, Vipera aspis</td></tr>
<tr><td colspan="2" align="center">Scientific classification</td></tr>
<tr><td>Kingdom:</td><td>Animalia</td></tr>
<tr><td>Phylum:</td><td>Chordata</td></tr>
<tr><td>Subphylum:</td><td>Vertebrata</td></tr>
<tr><td>Class:</td><td>Reptilia</td></tr>
<tr><td>Order:</td><td>Squamata</td></tr>
<tr><td>Suborder:</td><td>Serpentes</td></tr>
<tr><td>Family:</td><td>Viperidae</td></tr>
<tr><td>Subfamily:</td><td>Viperinae
Oppel, 1811</td></tr>
<tr><td colspan="2" align="center">Synonyms</td></tr>
<tr><td colspan="2">

- Viperini - Oppel, 1811
- Viperes - Cuvier, 1817
- Viperides - Latreille, 1825
- Viperina - Gray, 1825
- Viperiodea - Fitzinger, 1826
- Viperiodei - Eichwald, 1831
- Viperinae - Cantor, 1847
- Viperiformes - Günther, 1864
- Viperida - Strauch, 1869
- Atherini - Broadley, 1996[1]

</td></tr>
</table>

Common names*: pitless vipers, true vipers, Old World vipers,*[2] *true adders.*[3]

The **Viperinae**, or viperines, are a subfamily of venomous vipers found in Europe, Asia and Africa. They are distinguished by their lack of the heat-sensing pit organs that characterize their sister group, the Crotalinae. Currently, 12 genera and 66 species are recognized.[4] Most are tropical and subtropical, although one species, *Vipera berus*, even occurs within the Arctic Circle.[2]

Description

Members of this subfamily range in size from *Bitis schneideri*, that grows to a maximum of 28 cm, to *Bitis gabonica* that reaches a maximum length of over 2 m. Most species are terrestrial, but a few, such as *Atheris*, are completely arboreal.[2]

Although the heat-sensing pits that characterize the Crotalinae are clearly lacking in the viperines, a supernasal sac with sensory function has been described in a number of species. This sac is an invagination of the skin between the supranasal and nasal scales and is connected to the ophthalmic branch of the trigeminal nerve. The nerve endings here resemble those in the labial pits of boas. The supernasal sac is present in *Daboia*, *Pseudocerastes* and *Causus*, but is especially well developed in *Bitis*. Experiments have shown that strikes are not only guided by visual and chemical cues, but also by heat, with warmer targets being struck more frequently than colder ones.[2]

Geographic range

Europe, Asia and Africa.[1] However, they do not occur in Madagascar.[5]

Reproduction

Generally, members of this subfamily are viviparous (ovoviviparous), although a few, such as *Pseudocerastes*, lay eggs.[2]

Genera

Genus[4]	Taxon author[4]	Species[4]	Subsp.*[4]	Common name[2] [6]	Geographic range[1]
Adenorhinos	Loveridge, 1930	1	0	Uzungwe viper	Central Tanzania: Udzungwe and Ukinga Mountains.
Atheris	Cope, 1862	8	1	Bush vipers	Tropical subsaharan Africa, excluding southern Africa.
Bitis	Gray, 1842	14	2	Puff adders	Africa and the southern Arabian Peninsula.
Cerastes	Laurenti, 1768	3	0	Horned vipers	North Africa eastward through Arabia and Iran.
Daboia	Gray, 1842	1	1	Russell's viper	Pakistan, India, Sri Lanka, Bangladesh, Nepal, Myanmar, Thailand, Cambodia, China (Kwangsi and Kwangtung), Taiwan and Indonesia (Endeh, Flores, east Java, Komodo, Lomblen Islands).
Echis	Merrem, 1820	8	6	Saw-scaled vipers	India and Sri Lanka, parts of the Middle East and Africa north of the equator.
Eristicophis	Alcock and Finn, 1897	1	0	McMahon's viper	The desert region of Balochistan near the Iran-Afghanistan-Pakistan border.
Macrovipera	Reuss, 1927	4	4	Large Palearctic vipers	Semideserts and steppes of northern Africa, the Near and Middle East, and the Milos Archipelago in the Aegean Sea.
Montatheris	Boulenger, 1910	1	0	Kenya mountain viper	Kenya: moorlands of the Aberdare range and Mount Kenya above 3000 m.
Proatheris	Peters, 1854	1	0	Lowland viper	Floodplains from southern Tanzania (northern end of Lake Malawi) through Malawi to near Beira, central Mozambique.
Pseudocerastes	Boulenger, 1896	1	1	False horned viper	From the Sinai of Egypt eastward to Pakistan.

| *Vipera*[T] | Laurenti, 1768 | 23 | 12 | Palearctic vipers | Great Britain and nearly all of continental Europe across the Arctic Circle and on some islands in the Mediterranean (Elba, Montecristo, Sicily) and Aegean Sea eastward across northern Asia to Sakhalin Island and North Korea. Also found in northern Africa in Morocco, Algeria and Tunisia. |

*) Not including the nominate subspecies.

[T]) Type genus.

Taxonomy

Until relatively recently, two other genera were also included in the Viperinae. However, they were eventually considered so distinctive within the Viperidae, that separate subfamilies were created for them:[1]

- Genus *Azemiops* - moved to subfamily *Azemiopinae* by Liem, Marx & Rabb (1971).
- Genus *Causus* - recognition of subfamily *Causinae* (Cope, 1860) was proposed by Groombridge (1987) and further supported by Cadle (1992).

Nevertheless, these groups, together with the genera currently recognized as belonging to the Viperinae, are still often referred to collectively as the true vipers.[2]

Broadley (1996) recognized a new tribe, Atherini, for the genera *Atheris*, *Adenorhinos*, *Montatheris* and *Proatheris*, the type genus for which is *Atheris*.[1]

See also

- List of viperine species and subspecies
- Viperinae by common name
- Viperinae by taxonomic synonyms
- Snakebite

References

[1] McDiarmid RW, Campbell JA, Touré T. 1999. Snake Species of the World: A Taxonomic and Geographic Reference, vol. 1. Herpetologists' League. 511 pp. ISBN 1-893777-00-6 (series). ISBN 1-893777-01-4 (volume).

[2] Mallow D, Ludwig D, Nilson G. 2003. True Vipers: Natural History and Toxinology of Old World Vipers. Krieger Publishing Company, Malabar, Florida. 359 pp. ISBN 0-89464-877-2.

[3] U.S. Navy. 1991. Poisonous Snakes of the World. US Govt. New York: Dover Publications Inc. 203 pp. ISBN 0-486-26629-X.

[4] "Viperinae" (http://www.itis.gov/servlet/SingleRpt/SingleRpt?search_topic=TSN&search_value=563898). Integrated Taxonomic Information System. . Retrieved 4 August 2006.

[5] Stidworthy J. 1974. Snakes of the World. Grosset & Dunlap Inc. 160 pp. ISBN 0-448-11856-4.

[6] Spawls S, Branch B. 1995. The Dangerous Snakes of Africa. Ralph Curtis Books. Dubai: Oriental Press. 192 pp. ISBN 0-88359-029-8.

Further reading

- Breidenbach CH. 1990. Thermal cues influence strikes in pitless vipers. Journal of Herpetology 4: 448-50.
- Broadley DG. 1996. A review of the tribe Atherini (Serpentes: Viperidae), with the descriptions of two new genera. African Journal of Herpetology 45(2): 40-48.
- Cantor TE. 1847. Catalogue of reptiles inhabiting the Malayan Peninsula and Islands. Journal of the Asiatic Society of Bengal. Calcutta. 16(2): 607-656, 897-952, 1026-1078[1040].
- Cuvier G. 1817. Le règne animal distribué d'après son organisation, pour servir de base à l'histoire naturelle des animaux det d'introduction à l'anatomie comparée. Tome II, contenant les reptiles, les poissons, les mollusques et les annélidés. Déterville, Paris. xviii, 532 pp.[80].
- Eichwald, E. 1831. Zoologia specialis, quam expositis animalibus tum vivis, tum fossilibus potissimuni rossiae in universum, et poloniae in specie, in usum lectionum publicarum in Universitate Caesarea Vilnensi. Zawadski,

Vilnae. 3: 404 pp.[371].

- Fitzinger LJFJ. 1826. Neue classification der reptilien nach ihren natürlichen verwandtschaften. Nebst einer verwandtschafts-tafel und einem verzeichnisse der reptilien-sammlung des K. K. zoologischen museum's zu Wien. J.G. Hübner, Wien. vii, 66 pp.[11].
- Gray JE. 1825. A synopsis of the genera of reptiles and Amphibia, with a description of some new species. Annals of Philosophy, new ser., 10: 193-217[205].
- Günther ACLG. 1864. The Reptiles of British India. Ray Society. London. xxvii. 452 pp.[383].
- Latreille PA. 1825. Familles naturelles du règne animal, exposés succinctement et dans un ordre analytique, avec l'indication de leurs genres. Bailliere, Paris. 570 pp.[102].
- Lynn WG. 1931. The structure and function of the facial pit of the pit vipers. American Journal of Anatomy 49: 97.
- Oppel M. 1811. Mémoire sur la classification des reptiles. Ordre II. Reptiles à écailles. Section II. Ophidiens. Annales du Musée National d'Histoire Naturelle, Paris 16: 254-295, 376-393.[376, 378, 389].
- Strauch A. 1869. Mémoires de l'Academie des Sciences Impérial de St. Pétersbourg (7)14: 144 pp.[19]

Echis coloratus

Echis coloratus	

Scientific classification	
Kingdom:	Animalia
Phylum:	Chordata
Subphylum:	Vertebrata
Class:	Reptilia
Order:	Squamata
Suborder:	Serpentes
Family:	Viperidae
Subfamily:	Viperinae
Genus:	*Echis*
Species:	*E. coloratus*

Binomial name
Echis coloratus Günther, 1878

Synonyms
• *Echis froenata* - Duméril, Bibron & Duméril, 1854 • E[*chis*]. *carinata* var. *frenata* - Jan, 1863 • *Echis colorata* - Günther, 1878 • *Echis frenata* - Pfeffer, 1893 • *Echis coloratus* - Boulenger, 1896 • *Echis coloratus coloratus* - Cherlin, 1983 • *Echis* [(*Turanechis*)] *froenatus* - Cherlin, 1990[1]

Common names: *painted saw-scaled viper,*[2] *painted carpet viper, Burton's carpet viper,*[3] *more.*

Echis coloratus is a venomous viper species found in the Middle East and Egypt.[1] No subspecies are currently recognized.[4]

Description

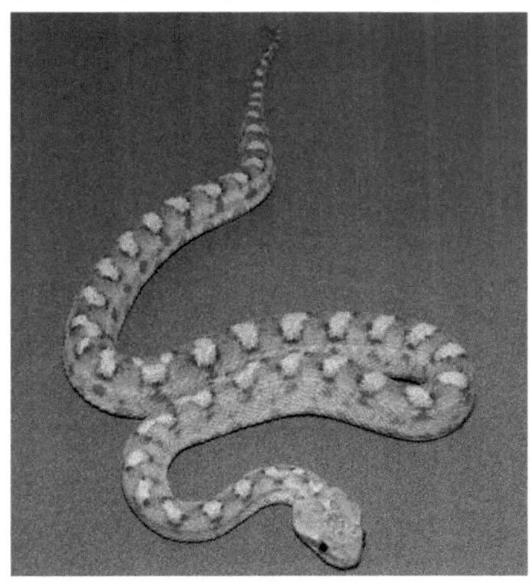

E. coloratus.

Grows to a maximum length of 75 cm.[2]

Common names

Painted saw-scaled viper,[2] painted carpet viper, Burton's carpet viper,[3] Palestine saw-scaled viper,[5] Arabian saw-scaled viper,[6] Mid-East saw-scaled viper.[6]

Geographic range

Found in the Middle East in Sinai, Israel and Jordan. On the Arabian Peninsula: Saudi Arabia, Yemen and Oman. In Africa it occurs in eastern Egypt east of the Nile and as far south as the 24th parallel. The type locality given is "Jebel Shárr, at an altitude of 4500 feet ... Midian" (Saudi Arabia, 1371 m altitude).[1]

Habitat

Occurs in rocky deserts, from sea level to altitudes as high as 2,500 m. Not found in sandy deserts.[3]

Taxonomy

In order to maintain nomenclatural stability, Stimson (1974) proposed that *E. coloratus* be validated over *E. froenata*. The ICZN subsequently gave *coloratus* precedence over *froenata* by use of its plenary powers.[1]

See also

- List of viperine species and subspecies
- Viperinae by common name
- Viperinae by taxonomic synonyms
- Snakebite

References

[1] McDiarmid RW, Campbell JA, Touré T. 1999. Snake Species of the World: A Taxonomic and Geographic Reference, vol. 1. Herpetologists' League. 511 pp. ISBN 1-893777-00-6 (series). ISBN 1-893777-01-4 (volume).

[2] Mallow D, Ludwig D, Nilson G. 2003. True Vipers: Natural History and Toxinology of Old World Vipers. Krieger Publishing Company, Malabar, Florida. 359 pp. ISBN 0-89464-877-2.

[3] Spawls S, Branch B. 1995. The Dangerous Snakes of Africa. Ralph Curtis Books. Dubai: Oriental Press. 192 pp. ISBN 0-88359-029-8.

[4] "*Echis coloratus*" (http://www.itis.gov/servlet/SingleRpt/SingleRpt?search_topic=TSN&search_value=634968). Integrated Taxonomic Information System. . Retrieved 1 August 2006.

[5] *Echis coloratus* (http://reptile-database.reptarium.cz/species.php?genus=Echis&species=coloratus) at the Reptarium.cz Reptile Database (http://reptile-database.reptarium.cz/). Accessed 2 August 2007.

[6] *Echis coloratus* (http://www.toxinfo.org/antivenoms/indication/ECHIS_COLORATUS.html) at Munich AntiVenom INdex (http://www.toxinfo.org/antivenoms/). Accessed 2 August 2007.

External links

- *Echis coloratus* (http://calphotos.berkeley.edu/cgi/img_query?stat=BROWSE& query_src=photos_fauna_sci-Reptile&where-lifeform=Reptile&where-taxon=Echis+coloratus) at CalPhotos (http://calphotos.berkeley.edu/). Accessed 2 August 2007.
- *Echis coloratus coloratus* Günther, 1878 (http://www.megasphera.cz/africanvenomoussnakes/_private/ Monografie_Echis_coloratus_coloratus.htm) at Tomáš Mazuch (http://www.megasphera.cz/ africanvenomoussnakes/). Accessed 24 November 2007.
- *Echis coloratus terraesanctae* Babocsay, 2003 (http://www.megasphera.cz/africanvenomoussnakes/_private/ Monografie_Echis_coloratus_terraesanctae.htm) at Tomáš Mazuch (http://www.megasphera.cz/ africanvenomoussnakes/). Accessed 24 November 2007.
- *Echis omanensis* Babocsay, 2004 (http://www.megasphera.cz/africanvenomoussnakes/_private/ Monografie_Echis_omanensis.htm) at Tomáš Mazuch (http://www.megasphera.cz/africanvenomoussnakes/). Accessed 24 November 2007.
- *Echis khosatzkii* Cherlin, 1990 (http://www.megasphera.cz/africanvenomoussnakes/_private/ Monografie_Echis_khosatzkii.htm) at Tomáš Mazuch (http://www.megasphera.cz/africanvenomoussnakes/). Accessed 24 November 2007.
- "What Fiery Flying Serpent Symbolized Christ?" (http://www.meridianmagazine.com/sci_rel/000609serpent. html) at Meridian Magazine (http://www.meridianmagazine.com). Accessed 16 December 2008.

Echis carinatus

<table>
<tr><td colspan="2" align="center">Echis carinatus</td></tr>
<tr><td colspan="2" align="center"></td></tr>
<tr><td colspan="2" align="center">Scientific classification</td></tr>
<tr><td>Kingdom:</td><td>Animalia</td></tr>
<tr><td>Phylum:</td><td>Chordata</td></tr>
<tr><td>Subphylum:</td><td>Vertebrata</td></tr>
<tr><td>Class:</td><td>Reptilia</td></tr>
<tr><td>Order:</td><td>Squamata</td></tr>
<tr><td>Suborder:</td><td>Serpentes</td></tr>
<tr><td>Family:</td><td>Viperidae</td></tr>
<tr><td>Subfamily:</td><td>Viperinae</td></tr>
<tr><td>Genus:</td><td>Echis</td></tr>
<tr><td>Species:</td><td>E. carinatus</td></tr>
<tr><td colspan="2" align="center">Binomial name</td></tr>
<tr><td colspan="2" align="center">Echis carinatus
(Schneider, 1801)</td></tr>
<tr><td colspan="2" align="center">Synonyms</td></tr>
<tr><td colspan="2">

- [*Pseudoboa*] *Carinata* - Schneider, 1801
- *Boa Horatta* - Shaw, 1802
- *Scytale bizonatus* - Daudin, 1803
- [*Vipera (Echis)*] *carinata* - Merrem, 1820
- [*Echis*] *zic zac* - Gray, 1825
- *Boa horatta* - Gray, 1825
- *Echis carinata* - Wagler, 1830
- *Vipera echis* - Schlegel, 1837
- *Echis (Echis) carinata* - Gray, 1849
- *Echis ziczic* - Gray, 1849
- *V*[*ipera*]. *noratta* - Jerdon, 1854
- *V*[*ipera (Echis*]. *carinata* - Jan, 1859
- *Vipera (Echis) superciliosa* - Jan, 1859
- *E*[*chis*]. *superciliosa* - Jan, 1863
- *Vipera Echis Carinata* - Higgins, 1873
- *Echis carinatus* - Boulenger, 1896

</td></tr>
</table>

- *Echis carinata* var. *nigrosincta* - Ingoldby, 1923 (nomen nudum)
- *Echis carinatus carinatus* - Constable, 1949
- *Echis carinatus* - Mertens, 1969
- *Echis carinatus* - Latifi, 1978
- *Echis* [(*Echis*)] *carinatus carinatus* - Cherlin, 1990
- *Echis carinata carinata* - Das, 1996[1]

Common names: *saw-scaled viper,*[2] *Indian saw-scaled viper, little Indian viper,* [3] *more.*

Echis carinatus is a venomous viper species found in parts of the Middle East and Central Asia, and especially the Indian subcontinent. It is the smallest member of the Big Four snakes. The Saw-scaled viper is one of the species which are responsible for causing the most snakebite cases and deaths due to various factors such as their frequently occurrences in high populated regions and their inconspicuousness nature. [4] Five subspecies are currently recognized, including the nominate subspecies described here.[5]

Description

E. c. carinatus, southern India.

Size ranges between 38 and 80 cm in length, but usually no more than 60 cm.[2]

Head distinct from neck, snout very short and rounded. The nostril between three shields and head covered with small keeled scales, among which an enlarged supraocular is sometimes present. There are 9-14 interocular scales across the top of the head and 14-21 circumorbital scales. 1-3 rows of scales separate the eye and the supralabials. There are 10-12 supralabials, the fourth usually largest, and 10-13 sublabials.[2] [6]

Midbody there are 25-39 rows of dorsal scales that are keeled scales with apical pits; on the flanks, these have serrated keels. There are 143-189 ventral scales that are rounded and cover the full width of the belly. The subcaudals are undivided and number 21-52, and the anal scale is single.[2] [6]

The color-pattern consists of a pale buff, grayish, reddish, olive or pale brown ground color, overlaid middorsally with a series of variably colored, but mostly whitish spots, edged with dark brown, and separated by lighter interblotch patches. A series of white bows run dorsolaterally. The top of the head has a whitish cruciform or trident pattern and there is a faint stripe running from the eye to the angle of the jaw. The belly is whitish to pinkish, uniform in color or with brown dots that are either faint or distinct.[2] [6]

Common names

- English - saw-scaled viper,[2] Indian saw-scaled viper, little Indian viper.[3]
- Sinhala - *vali polonga.*[7]
- Oriya - *Dhuli Naga.*[8]
- Pushtu - *phissi.*[8]
- Tamil - *surattai pambu.*[7] *viriyan pamboo, surutai vireyan*[8]
- Telugu - *thrachu paamu*
- Sindhi - *kuppur, janndi.*[8]
- Marathi - *phoorsa.*[8]
- Kannada - *kallu have.*[8]
- Malayalam - *anali*[8]
- Gujarati - *tarachha, zeri padkoo udaneyn.*[8]
- Hindi - *afai.*[8]
- Russian - *эфа песчаная*[9]
- [Iraq]- Said Dekhil snake[10] حية سيد دخيل

Geographic range

Asia. On the Indian subcontinent: India, Sri Lanka Bangladesh and Pakistan (including Urak near Quetta and Astola Island off the coast of Makran). In the Middle East: Oman, Masirah (Island), eastern United Arab Emirates, Iraq and southwestern Iran. In Central Asia: Afghanistan, Uzbekistan, Turkmenistan and Tadzikhistan. The type locality was not included in the original description; given as "Arni" (India) by Russell (1796:3).[1]

There are also reports that this species occurs in Iraq.[11] [12] It is found in Thiqar and Kirkuk governorates[13]

Habitat

Found on a range of different substrates, including sand, rock, soft soil and in scrublands. Often found hiding under loose rocks. Specimens have also been found in Balochistan at altitudes of up to 1982 m.[2]

Behavior

This species is mostly crepuscular and nocturnal, although there have been reports of activity during daylight hours.[2] During the daytime they hide in all kinds of places, such as deep mammal burrows, rock fissures and fallen rotted logs. In sandy environments, they may bury themselves leaving only the head exposed. Often, they are most active after rains or on humid nights.[14] This species is often found climbing in bushes and shrubs, sometimes as much as 2 m above the ground. When it rains, up to 80% of the adult population will climb into bushes and trees. Once, it was observed how some 20 individuals had massed on top of a single cactus or small shrub.[2]

E. carinatus sidewinding.

They are one of the species responsible for causing the most snakebite cases due to their inconspicuous nature. Its characteristic pose, a double coil with a figure of eight, with the head poised in the center, permits it to lash out like a released spring.[8]

They move about mainly by sidewinding: a method at which they are considerably proficient and alarmingly quick. They are also capable of other forms of locomotion, but sidewinding seems to be best suited to moving about in their usual sandy habitats. It may also keep them from overheating too quickly, as there are only two points of contact with the hot surface in this form of locomotion.[2]

In the northern parts of its range, these snakes hibernate in winter.[8]

Feeding

It feeds on rodents, lizards, frogs, and a variety of arthropods, such as scorpions, centipedes and large insects.[8] Diet may be varied according to availability of prey. High populations in some areas may be due to this generalist diet.[2]

Reproduction

The population in India is ovoviviparous. In northern India, mating takes place in the winter with live young being born from April through August. Occasionally, births have also been recorded in other months. A litter usually consists of 3 to 15 young that are 115-152 mm in length.[8] Mallow *et al.* (2003) mention a maximum litter size of 23.[2]

Venom

This species produces on the average of about 18 mg of dry venom by weight, with a recorded maximum of 72 mg. It may inject as much as 12 mg, whereas the lethal dose for an adult is estimated to be only 5 mg.[8] Envenomation results in local symptoms as well as severe systemic symptoms that may prove fatal. Local symptoms include swelling and pain, which appear within minutes of a bite. In very bad cases the swelling may extend up the entire affected limb within 12-24 hours and blisters form on the skin.[15] The venom yield from individual specimens varies considerably, as does the quantity injected per bite. The mortality rate from their bites is about 20%, and due to the availability of the anti-venom, deaths are much rare presently.[8]

Of the more dangerous systemic symptoms, hemorrhage and coagulation defects are the most striking. Hematemesis, melena, hemoptysis, hematuria and epistaxis also occur and may lead to hypovolemic shock. Almost all patients develop oliguria or anuria within a few hours to as late as 6 days post bite. In some cases, kidney dialysis is necessary due to acute renal failure (ARF), but this is not often caused by hypotension. It is more often the result of intravascular hemolysis, which occurs in about half of all cases. In other cases, ARF is often caused by disseminated intravascular coagulation.[15]

In any case, antivenin therapy and intravenous hydration within hours of the bite are vital for survival.[15] At least eight different polyvalent and monovalent antivenins are available against bites from this species.[3]

The venom from this species is used in the manufacture of several drugs. One is called *echistatin*, which is an anticoagulant. Even though many other snake venoms contain similar toxins, *echistatin* is not only especially potent, but also simplistic in structure, which makes it easier to replicate. Indeed, it is obtained not only through the purification of whole venom,[16] but also as a product of chemical synthesis.[17] [18] Another drug made from *E. carinatus* venom is called *ecarin* and is the primary reagent in the ecarin clotting time (ECT) test, which is used to monitor anticoagulation during treatment with hirudin.[19] [20] Yet another drug produced from *E. carinatus* venom is Aggrastat (Tirofiban).

Subspecies

Subspecies[5]	Taxon author[5]	Common name	Geographic range[2]
E. c. astolae	Mertens, 1970	Astola saw-scaled viper	Pakistan (Astola Island).
E. c. carinatus	(Schneider, 1801)	South Indian saw-scaled viper[21]	Peninsular India.
E. c. multisquamatus	Cherlin, 1981	Multiscale saw-scaled viper	From Uzbekistan to Iran in the south and east to western Pakistan.
E. c. sinhaleyus	Deraniyagala, 1951	Sri Lankan saw-scaled viper	Sri Lanka.
E. c. sochureki	Stemmler, 1969	Sochurek's saw-scaled viper	Southern Afghanistan, Pakistan, northern India, southern and central Iran, Oman and UAE.

See also

* List of viperine species and subspecies
* Viperinae by common name
* Viperinae by taxonomic synonyms
* Snakebite
* Sidewinding

References

[1] McDiarmid RW, Campbell JA, Touré T. 1999. Snake Species of the World: A Taxonomic and Geographic Reference, vol. 1. Herpetologists' League. 511 pp. ISBN 1-893777-00-6 (series). ISBN 1-893777-01-4 (volume).

[2] Mallow D, Ludwig D, Nilson G. 2003. True Vipers: Natural History and Toxinology of Old World Vipers. Krieger Publishing Company, Malabar, Florida. 359 pp. ISBN 0-89464-877-2.

[3] *Echis carinatus* antivenoms (http://www.toxinfo.org/antivenoms/indication/ECHIS_CARINATUS.html) at Munich Antivenom Index (http://www.toxinfo.org/antivenoms/). Accessed 13 September 2006.

[4] Whitaker Z. 1990. Snakeman. Penguin Books Ltd. 192 pp. ISBN 0-14-014308-4.

[5] "*Echis carinatus*" (http://www.itis.gov/servlet/SingleRpt/SingleRpt?search_topic=TSN&search_value=634967). Integrated Taxonomic Information System. . Retrieved 1 August 2006.

[6] Boulenger GA. 1890. The Fauna of British India, Including Ceylon and Burma. Reptilia and Batrachia. Taylor & Francis, London, xviii, 541 pp.

[7] Checklists of the Snakes of Sri Lanka (http://www.slwcs.org/facts/snakes.html) at the Sri Lanka Wildlife Conservation Society (http://www.slwcs.org/). Accessed 15 August 2007.

[8] Daniels,J. C. (2002) The Book of Indian Reptiles and Amphibians, BNHS & Oxford University Press, Mumbai, pp 151-153. ISBN 019-566099-4

[9] эфа песчаная (http://www.floranimal.ru/pages/animal/je/1569.html) at Floranimal.ru (http://www.floranimal.ru/). Accessed 21 September 2008.

[10] http://www.uobaghdad.edu.iq/ArticleShow.aspx?ID=384

[11] Snakes and Spiders (http://www.blackfive.net/main/2006/06/snakes_and_spid.html) at Black Five (http://www.blackfive.net/). Accessed 6 January 2007.

[12] Joint Chiefs of Staff Campaign Analysis Report, Snakes & Scorpions in Iraq & Antivenin Sources.pdf Venomous Snakes and Scorpions in Iraq, and Their Antivenin Sources (http://www.brooks.af.mil/web/af/courses/amp/cluebag/Venomous) at 311th Human Systems Wing, Brooks City-Base (http://www.brooks.af.mil/). Accessed 6 January 2007.

[13] http://www.alforattv.net/index.php?show=news&action=article&id=58298

[14] Mehrtens JM. 1987. Living Snakes of the World in Color. New York: Sterling Publishers. 480 pp. ISBN 0-8069-6460-X.

[15] Ali G, Kak M, Kumar M, Bali SK, Tak SI, Hassan G, Wadhwa MB. 2004. Acute renal failure following echis carinatus (saw–scaled viper) envenomation. Indian Journal of Nephrology 14:177-181. PDF (http://medind.nic.in/iav/t04/i4/iavt04i4p177.pdf#search=""echis carinatus" +envenomation") at Indian Medlars Centre (http://medind.nic.in/). Accessed 12 September 2006.

[16] Echistatin from *Echis carinatus* (https://www.sigmaaldrich.com/sigma/datasheet/E1518dat.pdf) at Sigma-Aldrich. Accessed 29 September 2006.

[17] Saw-scaled Vipers (http://www.portfolio.mvm.ed.ac.uk/studentwebs/session2/group13/vipers.html) at Electronic Medical Curriculum (http://www.portfolio.mvm.ed.ac.uk/). Accessed 29 September 2006.

[18] Garsky VM, Lumma PK, Freidinger RM, Pitzenberger SM, Randall WC, Veber DF, Gould RJ, Friedman PA. 1989. Chemical synthesis of echistatin, a potent inhibitor of platelet aggregation from Echis carinatus: synthesis and biological activity of selected analogs. USA: Proc. Natl. Acad. Sci. Vol.86(11):4022–4026. PDF (http://www.pubmedcentral.nih.gov/picrender.fcgi?artid=287380&blobtype=pdf) at PubMed Central (http://www.pubmedcentral.nih.gov/). Accessed 29 September 2006.

[19] Fabrizio MC. 2001. Use of Ecarin Clotting Time (ECT) with Lepirudin Therapy in Heparin-Induced Thrombocytopenia and Cardiopulmonary Bypass. JECT 33:117–125. PDF (http://www.amsect.org/ce/HIT/ject_2001_v33_n2_fabrizio.pdf) at Journal of The American Society of ExtraCorporeal Technology (http://www.amsect.org/). Accessed 5 June 2007.

[20] Textarin/Ecarin Time (http://www.specialtylabs.com/books/display.asp?id=1060) at Specialty Laboratories (http://www.specialtylabs.com/). Accessed 5 June 2007.

[21] Checklist of Indian Snakes with English Common Names (http://www.bio.utexas.edu/grad/sp/pubs/Indian Snakes-Checklist.pdf) at University of Texas (http://www.biosci.utexas.edu/). Accessed 22 October 2006.

Further reading

- Hughes, B. 1976 Notes on African carpet vipers, Echis carinatus, Echis leucogaster and Echis ocellatus (Viperidae, Serpentes). Rev. suisse Zool. 83 (2): 359-371.
- Schneider JG. 1801. Historiae Amphibiorum naturalis et literariae. Fasciculus secundus continens Crocodilos, Scincos, Chamaesauras, Boas. Pseudoboas, Elapes, Angues. Amphisbaenas et Caecilias. Frommani, Jena. 364 pp.

External links

- *Echis carinatus* (http://reptile-database.reptarium.cz/species.php?genus=Echis&species=carinatus) at the Reptarium.cz Reptile Database (http://reptile-database.reptarium.cz/). Accessed 2 August 2007.
- *Echis carinatus* (image 1) (http://itgmv1.fzk.de/www/itg/uetz/herp/photos/Echis_carinatus1.jpg) at the Institute of Toxicology and Genetics (http://itgmv1.fzk.de/). Accessed 12 September 2006.
- *Echis carinatus* (image 2) (http://itgmv1.fzk.de/www/itg/uetz/herp/photos/Echis_carinatus2.jpg) at the Institute of Toxicology and Genetics (http://itgmv1.fzk.de/). Accessed 12 September 2006.
- *Echis carinatus* (image 3) (http://itgmv1.fzk.de/www/itg/uetz/herp/photos/Echis_carinatus3.jpg) at the Institute of Toxicology and Genetics (http://itgmv1.fzk.de/). Accessed 12 September 2006.
- *Echis carinatus* (http://members.fortunecity.com/ukp001/naja/viperidae/echis_carinatus.htm) at Snakes of Sri Lanka (http://members.fortunecity.com/ukp001/naja/mainpage.htm). Accessed 22 October 2006.
- *Echis carinatus carinatus* (Schneider, 1801) (http://www.megasphera.cz/africanvenomoussnakes/_private/Monografie_Echis_carinatus_carinatus.htm) at Tomáš Mazuch (http://www.megasphera.cz/africanvenomoussnakes/). Accessed 24 November 2007.

Article Sources and Contributors

Vipera_ammodytes_gregorwallneri *Source*: http://en.wikipedia.org/w/index.php?title=Vipera_ammodytes_gregorwallneri *Contributors*: Droll, Jwinius, Lightmouse, Mr. Lefty, Tim Q. Wells

List_of_viperine_species_and_subspecies *Source*: http://en.wikipedia.org/w/index.php?title=List_of_viperine_species_and_subspecies *Contributors*: Droll, Dsmdgold, Fayenatic london, Jwinius, Lightmouse, That Guy, From That Show!, Tirkfl, 12 anonymous edits

Vipera ammodytes *Source*: http://en.wikipedia.org/w/index.php?title=Vipera_ammodytes *Contributors*: Angelo De La Paz, Arctosouros, BanyanTree, Branka France, Crnorizec, Dawson, Droll, Dysmorodrepanis, Eastunionlancer, Edgar181, Elenaterkel, Eupator, Feci1024, Filip Lazarevic, Gouerouz, HarisM, Heli12, IvanTortuga, Jwinius, Kaarel, Khatru2, Lightmouse, Marklar2007, Mikie2, Miqe, Nabokov, Navay, Nik666, Rich Farmbrough, Rigadoun, Sam Blacketer, Tauiota2, The Myotis, Tim Q. Wells, Ucucha, Wlodzimierz, Zarkus, 55 anonymous edits

Cerastes cerastes *Source*: http://en.wikipedia.org/w/index.php?title=Cerastes_cerastes *Contributors*: Alansohn, Bobby122, Bschott, CT Cooper, Cerastesgr, Da Wenis, Dawson, Dominic, Dreadstar, Droll, Dsmdgold, Edgar181, Felix Folio Secundus, Florian Huber, Frankyboy5, Gaius Cornelius, Grika, HCA, Hwoarang17, Insanity Incarnate, IvanTortuga, Jwinius, Karsten Kohls, La Pianista, Lightmouse, Mad Max, Mwng, Nabokov, Nachoman-au, PTSE, Prodego, Qwertyuiopaaronboog, ReyBrujo, Seco201, Sluzzelin, Staffwaterboy, SwisterTwister, Tacubus, The Rambling Man, Tim Q. Wells, Vincent shooter, Yath, Zahid Abdassabur, ص راوي مغربي, 48 anonymous edits

Vipera berus *Source*: http://en.wikipedia.org/w/index.php?title=Vipera_berus *Contributors*: Achowat, Adrian.benko, Alansohn, Andres, Arbitrary username, AshLin, B kimmel, Baronnet, BillC, Bjelleklang, Bob Burkhardt, Brandon Steven Piña Sanchez, Caissaca, CanisRufus, Casliber, Chuunen Baka, CoWiiEs, Crnorizec, Cygnis insignis, Darwinek, Dekisugi, Derek.cashman, Droll, Edgar181, Ericoides, Frietjes, Gdr, Ginkgo100, Gits (Neo), GoingBatty, Goodnightmush, Gouerouz, Graminophile, HamburgerRadio, Hapsiainen, Highlandtiercel, Hordaland, Hu12, Ialsoagree, IvanTortuga, Jaan513, Jaberwocky6669, Jauhienij, Jimfbleak, John Vandenberg, Jukkapietil, Jwinius, Kaarel, KevinOKeeffe, Khoikhoi, KimvdLinde, Kiwifruitrulz, KnowledgeOfSelf, Koavf, Kuru, Leszek Jańczuk, Leyo, Lightmouse, Lisapollison, Lokyz, Mad Max, Matta1304, Metanoid, Mgiganteus1, Minimac, Moon83, Moonriddengirl, Mthibault, Mwng, Nabokov, Nehrams2020, Ohconfucius, Ooh Vicar, Owllight, Pavel Vozenilek, Pcb21, Petter Bøckman, PhilKnight, Philip Trueman, Pigsonthewing, Pinethicket, Pinky sl, Pkuczynski, Pschemp, PythonParrot, Ranaboy, Random updater, Rich Farmbrough, Richard New Forest, Rjwilmsi, Rkarlsba, RonDivine, Saadee, Sam Hocevar, Samsara, Sebastian T Russell, Sebastian80, Senix, Shenme, Shoefly, Simonmaal, Snowmanradio, Sokac121, Sooohood, Sorstalan, Stemonitis, Surv1v4l1st, Tim Q. Wells, UW, YotePole, Zarkus, 117 anonymous edits

Viperinae *Source*: http://en.wikipedia.org/w/index.php?title=Viperinae *Contributors*: Aelfthrytha, AjaxSmack, AlimanRuna, Anduzan, Bob Burkhardt, Brockert, Canderson7, CanisRufus, Catbar, Cyp, DJ Clayworth, DabMachine, Dawson, DocWatson42, Droll, Esperant, Esprit15d, Estel, Feydey, Gdr, GerardM, Glenn, Gouerouz, GraemeL, Gökhan, Hbdragon88, Hongooi, Idleguy, Jwinius, Jwrosenzweig, Kchishol1970, Lightmouse, Linux using monkey, Marj Tiefert, Matt Crypto, Meelar, Mwng, Pcb21, Phil1988, Ravn, Rchamberlain, RedWolf, Rob Hooft, Rogper, RoyBoy, Ryanmcdaniel, S.K., Sadalmelik, Sam Hocevar, SamH, SchuminWeb, Seba, Shimgray, Stemonitis, Tabletop, TheAlphaWolf, Thesexualityofbereavement, Tim Q. Wells, TonyCB4, Unyoyega, UtherSRG, Val42, WojPob, Yath, Zoicon5, 58 anonymous edits

Echis coloratus *Source*: http://en.wikipedia.org/w/index.php?title=Echis_coloratus *Contributors*: Alro, Caissaca, CanisRufus, DanielCD, Droll, Dysmorodrepanis, Fuzheado, Gdr, Infrogmation, J.smith, Jcurtis, Jwinius, Kaarel, Mwng, Oliverkeenan, Pcb21, ReyBrujo, Stemonitis, Struhs, Tim Q. Wells, Tt100, Yath, 9 anonymous edits

Echis carinatus *Source*: http://en.wikipedia.org/w/index.php?title=Echis_carinatus *Contributors*: Aelfthrytha, Amit6, AshLin, Barticus88, Basawala, Beetstra, Chipmunkdavis, DRosenbach, Dpv, Droll, Dsmdgold, Fearingpredators, Ferritecore, Frankie816, Gaius Cornelius, Gdr, Gökhan, HCA, Idleguy, Jahangard, Jwinius, Kaarel, Karthickbala, Knuckles, Kummi, Lightmouse, Mad Max, Muralika, Mwng, Nabokov, Orlica, Pcb21, Pinethicket, RitigalaJayasena, ShelfSkewed, Shyamal, Thaurisil, Tim Q. Wells, Unstudmaddu, UtherSRG, Wikipedia crusader, Woohookitty, Yath, Александър, 34 anonymous edits

Image Sources, Licenses and Contributors

Printed by Books on Demand GmbH, Norderstedt / Germany